Process Development Division 2019

Held at the 2019 AIChE Annual Meeting

Orlando, Florida, USA
10 - 15 November 2019

ISBN: 978-1-7138-0535-9

Printed from e-media with permission by:

Curran Associates, Inc.
57 Morehouse Lane
Red Hook, NY 12571

AIChE
120 Wall Street, FL 23
New York, NY 10005-4020

Phone: (800) 242-4363
Fax: (203) 775-5177

www.aiche.org

Additional copies of this publication are available from:

Curran Associates, Inc.
57 Morehouse Lane
Red Hook, NY 12571 USA
Phone: 845-758-0400
Fax: 845-758-2633
Email: curran@proceedings.com
Web: www.proceedings.com

TABLE OF CONTENTS

571591 Demystifying Mass Transfer in Electrolyzers through Rapid Prototyping with 3D Printed Parallel Plate Cells

As a result of the global energy transition, a large percentage of the world's power supply will become electric. For electrochemical engineers this transition generates a sea of opportunities. After all, it is electrolyzers that use electricity to produce some of life's essential chemicals. However, fully embracing these opportunities in a sustainable way is challenging: Future electrolyzers must be safer, cleaner, and far more flexible, while simultaneously increasing productivity and cost-effectiveness. Essentially, they will have to achieve more with less.

One key parameter in the performance of electrolyzers is the rate of mass transfer. Most often, electrochemical reactions are mass transfer limited due to the fast kinetics that depend exponentially on the applied overpotential. Numerous correlations in literature exist to describe the rate of mass transfer in a specific reactor.[1] Though, their predictions can vary up to almost an order of magnitude (see figure 1). The reason for this is the large geometric variety between the reactors that were measured. The design of the inlets in particular is a crucial contributor to the rate of mass transfer. [2] Moreover, only a few lab-scale electrolyzers are commercially available, often with sub-optimal designs, which limits further research. As a consequence, it is difficult to establish a generalized description of the mass transfer performance of parallel plate electrolyzers. Furthermore, most existing correlations only describe the average rates of mass transfer and therefore do not consider any local variation despite the importance thereof.[3]

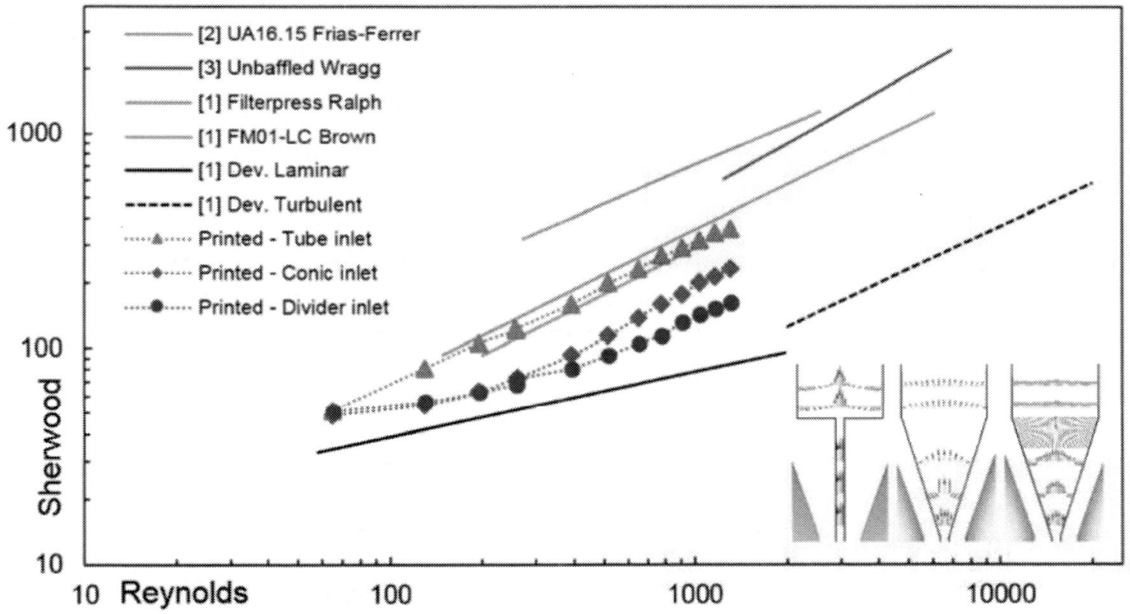

Figure 1: Mass transfer performance in terms of Sherwood vs. Reynolds. Up to an order of magnitude of difference in performance is visible between reactors. The red triangle, square and circle markers denote a selection of inlet manifolds tested with the 3D printed prototype. Small insert: sketches of the inlet manifolds, from left to right: tube inlet, conic inlet, divider-type inlet.

The goal of our study has been to demystify the fundamentals of mass transfer in parallel plate electrolyzers. As this would have required testing a large geometric variety of cells, traditional manufacturing techniques were impractical due to their lead time and cost. Therefore, we have used 3D printing to rapidly construct many different reactor prototypes. Moreover, each prototype could be finetuned to the exact specifications each measurement required. This way, each geometric parameter could be varied and investigated individually.

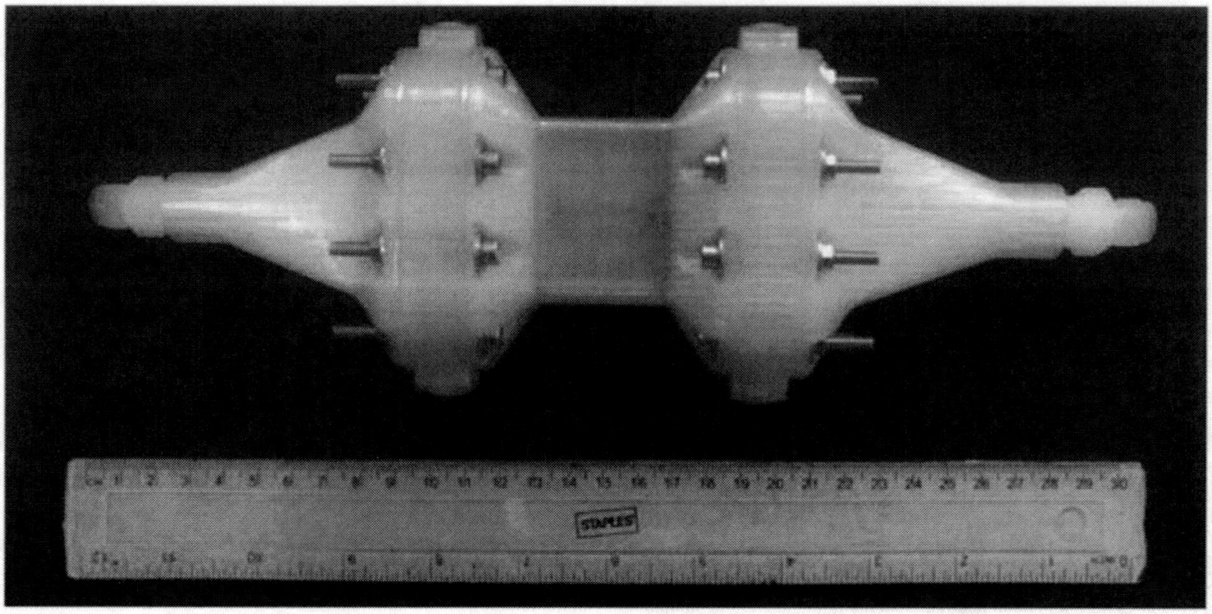

Figure 2: 3D printed electrolyzer prototype for mass transfer studies. The electrode area is 40 cm^2.

Nearly the entire reactor can be 3D printed in less than a workweek's time (see figure 2). Inlet manifolds, the channel and the rubber sealings could all be printed using commercially available equipment. Furthermore, by using 32 independently operated electrode segments we were able to measure the mass transfer performance on a local scale. Sherwood – Reynolds correlations were established for designs with different shapes of turbulence promoters, 3D electrodes and inlet manifolds. The effect of an increased or decreased hydrodynamic entrance length was also investigated.

In the end these prototypes lead to enhanced understanding of electrochemical reactors, allowing us to intensify processes such as alkaline water electrolysis. As a proof of concept, we are developing a 3D printed lab-scale electrolyzer for alkaline water splitting which can be used to test the effect of geometry on the production of hydrogen directly.

References

[1] T.R. Ralph, M.L. Hitchman, J.P. Millington, F.C. Walsh, *Mass Transport in an electrochemical laboratory filterpress reactor and its enhancement by turbulence promoters,* Electrochem. Acta, **1994,** pp. 591-603.

[2] A. Frías-Ferrer, J. Gonzálex-García, V.Sáez, C. Ponce de León, F.C. Walsh, *The effects of manifold flow on mass transport in electrochemical filter-press reactors,* AIChE J., **2008,** pp. 811-823.

[3] A.A. Wragg, A.A. Leontaritis, *Local mass transfer and current distribution in baffled and unbaffled parallel plate electrochemical reactors,* Chem. Eng. J., **1997,** pp. 1-10.

574364 Additive Manufacturing of Catalyst Support Structures with in Operando Adjustable Mass Transport and Flow Characteristics

In conventional randomly packed bed reactors, the flow and mass transport characteristics are predefined by the particle diameter and the tube diameter. Structured reactors offer the opportunity to tailor mass transport and flow field according to the needs of the reaction system. In this regard, additively manufactured so-called Periodic Open Cellular Structures (POCS) as catalyst supports have shown their advantages over conventional randomly packed bed reactors in terms of heat management and pressure drop [1, 2]. However, since the internal flow field, pressure drop and therefore the resulting mass transport characteristics are mainly influenced by the geometrical properties of the unit cell, for POCS the resulting transport characteristics of the structure is already determined and fixed by the choice and design of the unit cell. To achieve more flexibility, we have recently proposed a new type of so-called interpenetrating POCS (interPOCS) which are formed by inserting a diamond unit cell based structure in the cavities of a second diamond cell structure. These interPOCS allow for in operando flow field and mass transport adjustments by shifting the (relative) position of the loose structure [3].

With detailed computational fluid dynamics (CFD) based simulations with an additional in-house particle-tracking algorithm implementation we systematically investigated the flow field and mass transport characteristics within the interPOCS in dependency of the relative positioning of the loose structure. The highly adjustable structure allows for a broad variation of mass transport characteristics in a catalytic reactor without changing the system's periphery. This functionality enables mass transport optimization for a given system and represents a completely new application of additively manufactured catalyst support structures for highly flexible in operando tuning of flow field and mass transport characteristics.

References

[1] C. Busse, H. Freund, and W. Schwieger (2018). *Chem. Eng. Process.* 124, 199-214.

[2] A. Inayat, M. Klumpp, M. Lämmermann, H. Freund and W. Schwieger (2016). *Chem. Eng. J.* 287, 704-719.

[3] G. Do, T. Stiegler, M. Fiegl, L. Adler, C. Körner, A. Bösmann, H. Freund, W. Schwieger and P. Wasserscheid (2017). *Ind. Eng. Chem. Res.* 45, 13402-13410.

561588 Role of 3D Printed and LED-Driven Photostation in Photocatalytic and Photo-Fenton Activity of Iron Oxide Doped Graphitic Carbon Nitride

The degree of photocatalysis or any photolysis reaction depends on the intensity of light. In most cases, a major portion of light is not used as it does not reach the reaction zone thereby wasting a significant portion of light and energy, as well. In order to handle this situation, we have designed and 3D printed a photostation with complex parabolic reflective surface, so that all the incident light beams can reach the reaction zone. The photostation is fitted with two chips-on-board (COB) light emitting diodes (LEDs) reaching the total (visible) light intensity within the reaction zone over 290,000 lux (For comparison, it is commonly reported that the sun emits anywhere from 32,000 to 100,000 lux onto the Earth). Light simulation analysis confirmed that all incident light beams can reach the reaction zone of photostation. In order to test the performance of this photostation, we have used 0.5 to 2 wt.% iron oxide doped graphitic carbon nitride (g-C_3N_4) to degrade a model water pollutant, dicamba (3,6-Dichloro-2-methoxybenzoic acid), which is common herbicide, in presence of traces of hydrogen peroxide and visible light. It has been confirmed experimentally that the rate of degradation of dicamba is directly proportional to the light intensity and geometry of photostation thereby confirming the positive role of photostation design. The rapid degradation of dicamba under this study can be attributed to the four distinct phenomena of photocatalytic activity of carbon nitride, quenching of electron/hole pairs and generation of additional reactive hydroxyl radicals by hydrogen peroxide, Fenton and photo-Fenton activity of iron oxide component of carbon nitride in presence of hydrogen peroxide and photocatalytic activity of iron oxide alone in conjuncture with carbon nitride. Liquid Chromatography Mass spectrometry (LCMS) analysis of the degradation products showed loss of chlorine from the aromatic ring and evidence of free radical addition reactions in the course of photocatalysis.

557053 Solution-Based 3D Printing of Hierarchical Microporous Polymers for Energy-Efficient Adsorption

Current 3D printing methods have strong limitations in the classes of polymers that can be printed. This means that many polymers of technological interest cannot currently be 3D printed. Here, we demonstrate a generalizable method for 3D printing of viscoelastic polymer solutions and apply it to both porous (a polymer of intrinsic microporosity, PIM-1) and non-porous (e.g., cellulose acetate, Matrimid) polymers. Successful ternary ink formulations that result in the creation of three-dimensional structures with hierarchical porosity require balancing of solution thermodynamics (phase separation), mass transfer (solvent evaporation), and rheology. To demonstrate the utility of this approach, a microporous polymer (PIM-1) that cannot be printed with current 3D printing technologies is 3D printed into a high-efficiency respirator exhibiting hierarchical porosity ranging from sub-nanometer to millimeter pores. These short mass transfer contactors (1.27 cm) were challenged with a flow of toluene vapor (10000 ppm) in N_2 gas. Owing to the precisely designed gas distribution channels, the 3D printed respirator were found to create a pure (< 1 ppm toluene) N_2 stream for 1.7 hours, 6 times longer than traditional structures fabricated from the same PIM-1 material, and more than 4,000 times the residence time of gas in the bed. This solution-based additive manufacturing approach significantly extends the material spectrum of 3D printing technologies. By 3D printing adsorptive materials into precisely designed mass transfer contactor, the performance, energy efficiency, and material efficiency can be significantly improved.

559790 Titanium/Titania Electrocatalyst for Hydrogen Evolution Synthesized Via a Combination of 3D Printing and Anodization

A titanium core/titania coating, hierarchically structured catalyst was fabricated via combining 3D printing and in-situ anodization. 3D printing enabled control of the macro-porosity of the titanium core. Anodization was used in order to grow highly-ordered TiO_2 nanotubes (NTs) on the metal core. The dimensions of the TiO_2 nanotubes were controlled via tuning of anodization parameters. The combination of both fabrication techniques allows for hierarchical control of the catalyst structure.

The performance of the reported 3D Ti/TiO_2 electrocatalyst was tested for hydrogen-evolution reaction (HER) via water electrolysis. Titanium dioxide is a known photo-electro-catalysts for water splitting, but competitive electrocatalytic activity for HER in the absence of light has not been reported. Onset potentials for HER as low as -45 mV vs. RHE were obtained for our electrodes in acid medium, making them competitive with current state-of-the-art electrodes aiming to replace platinum. Tafel slopes between 40 and 52 mV dec^{-1} point towards a Volmer-Heyrovsky mechanism, with the Heyrovsky reaction as possible controlling step.

The proposed electrocatalysts feature the following advantages: (1) the fabrication method for the 3D Ti/TiO_2 catalyst is simple, scalable and can be readily incorporated as printed reactor components into different technologies; (2) the 3D Ti/TiO_2 catalyst features a flow-through architecture that not only enables more efficient utilization of the surface area of the catalyst but that could enable large-scale, continuous-flow hydrogen-production reactor technologies; and (3) minimization of electrical resistance between active surface area (TiO_2 NTs) and current collector.

556480 Advanced Manufacturing Enabling Process Intensification for Catalytic Processes

Advanced manufacturing is a transformative and disruptive approach to industrial production that enables the rapid synthesis of virtually any geometry composed of a wide range of materials, such as polymers, ceramics, and metals. Process intensification is a set of novel approaches to process engineering and equipment design that leads to substantially smaller, more selective, and energy-efficient processing plants. The integration of advanced manufacturing with process intensification accelerates process development and deployment through rapid prototyping and enables new hardware that overcomes transport limitations. Advanced manufacturing (ex. 3D printing or additive manufacturing) is the physical backbone of Industry 4.0 that can take advantage of the conclusions of big data and machine learning by allowing design freedom. This expansion of the material design window can be utilized to reduce GHG intensity for novel or existing processes.

Numerous conversion technologies in the energy industry suffer from rapid coke build-up and subsequent deactivation of a catalytic material that is exposed to a carburizing environment. The time scale of deactivation often determines the reactor technology and process configuration employed. Commercial examples of deactivating catalysts range from seconds in a fluid bed catalytic cracking unit to minutes for moving bed propane dehydrogenation and days for continuous catalytic reformers or even years for steam methane reformers. A common step for these processes is the regeneration of the catalytic material via combustion of the coke. However, the cyclic process of coking in a carburizing environment followed by combustion in an oxidizing environment can often be detrimental to catalyst performance.

This work will discuss how advanced manufacturing methods can enable construction of materials capable of rapidly switching between two different environments with enhanced catalytic activity. Such materials capable of withstanding cyclic carburizing and oxidizing steps can be efficiently integrated into novel reactor concepts. The conversion of light gas is an excellent probe reaction for elucidating this behavior. In an effort to search for robust materials, honeycomb structures composed of metal were fabricated by direct metal laser sintering (DMLS), a 3D printing technique that involves the layer-wise successive deposition of micron-sized powder followed by laser induced sintering. The cyclic operation that alternates between carburizing and oxidizing environments leads to compositional modification at the surface of the monolith channels that imparts in-situ catalytic activity.

579939 Data Science and Machine Learning in the Upstream Oil and Gas Industry

Digital technologies have radically transformed many industries in the last two decades, but advances in pervasive data access, machine learning, and digital business models have penetrated the upstream oil and gas industry more slowly. This process is now intensifying, with digital technologies poised to disrupt core business and technical workflows, in production operations, drilling, and subsurface characterization, among other areas. I will review some specific instances where digital technologies, especially those enabled by data science and machine learning, have been successfully utilized within ExxonMobil, drawing lessons for the effective development and deployment of these technologies to achieve transformational business impact.

579937 Big Data Analytics in the Advanced Manufacturing of Biopharmaceuticals

This presentation describes the use of process data analytics and machine learning to improve process productivity, reliability, and control in next-generation biopharmaceuticals manufacturing systems. New information streams are described, including the use of hyperspectral imaging to generate gigabytes of real-time data in the form of high order tensors, and the coupling of on-line automated sampling systems with liquid chromatography-mass spectroscopy (LC-MS) to simultaneously measure transient variations in large numbers of distinct small molecules and proteins during manufacturing. Experimental data are presented from multiple laboratories for a variety of bioprocess unit operations, including bioreactors and freeze-drying systems. The use of that data in first-principles, data-driven, and hybrid models are described, with relationships drawn to chemometrics (e.g., partial least squares, principal component analysis), system identification (e.g., canonical variate analysis), pattern recognition (e.g., Fisher and quadratic discriminant analysis), and machine learning (e.g., ensemble learning, elastic net regularization).

581227 Big Data Impact on the Future of Manufacturing

There is an opportunity for innovation in chemical and pharmaceutical manufacturing processes through the integration of human intuition and experience with big data, machine learning and other software innovations. Subject matter experts bring their experience and expertise for constructing an effective analytics strategy, whereas new capabilities in software enable them to work and analyze processes and outcomes at the speed of thought. This speed of insight is increasingly important in dealing with two of the highest priority pressure points in the industry: time to market from research to production, and replacing batch-based production with a continuous production model.

However, since analysis can suffer from human biases and heuristics, it is imperative to take advantage of big data platforms and applications to ensure completeness of the analytics effort. This includes data collection and aggregation for comprehensive access to production results, as well as advanced analytics to accelerate the time to insight and the breath of questions which may be required for insight. This session will provide a framework for leveraging software innovation in the context of industry requirements with examples of current customer deployments and insights.

580269 Data-Assisted Modeling and Optimization for Process Systems Engineering

Coupling equation-based first-principle modeling with information that comes in the form of input-output data for enhanced decision-making, is gaining significant interest in process systems engineering [1]. Most data-driven optimization techniques consider the system under study entirely as a "black-box", however, in order to take advantage of years of obtained physical process knowledge, this work proposes a hybrid modeling and optimization approach [2-3]. This talk will demonstrate the performance of a suite of flexible modeling and optimization tools that are able to incorporate both physics-based and machine-learning components, to enable reliable data-assisted decision making. Specifically, adaptive and tractable surrogate functions will be developed to represent input-output data, and these will be merged with mechanistic equations in the form of constraints or discrepancy modeling [4]. Optimization is performed via a custom-based branch-and-bound framework that incorporates sampling, machine learning and bounding of the uncertainty of the approximations. Finally, the performance of the proposed methods is demonstrated through a set of process design case studies that rely on multifidelity models and data.

References:

1. Venkatasubramanian, V. (2019), AIChE J., 65: 466-478
2. Boukouvala, F., Hasan, MMF, Floudas, C.A. (2017). JOGO, 67(1-2):3-42
3. Kim, S.H. and Boukouvala, F. (2019), Optimization Letters, 10.1007/s11590-019-01428-7
4. Kennedy, M. C. and O'Hagan, A. (2001), Journal of the Royal Statistical Society, 63: 425-464

575002 Model Based Plant-Wide Optimization of an Industrial Scale Continuous Tank Crystallizer Network for Dextrose Monohydrate Crystallization

This work is aimed to present a successful industrial-academic collaboration whose aim was to optimize a continuous large-scale dextrose crystallization network producing crystalline dextrose monohydrate. To put into the context, the throughput of the plant is in order of metric tons'/h solid crystals, whereas the combined nominal volume of the continuous tank crystallizers is in order of hundreds of m^3. The main quality attribute of the product is the fine index, which represents the volume fraction of crystals under 75 μm. The maximally allowed fine index is 25 %, and the average fine index of the product significantly exceeded this value. Hence, the goal of the work was to carry out a model based fine index minimization for the current operation as well as the technically feasible combinations of the available crystallizers, also including technically realizable recirculation streams. A significant challenge was the time restriction (order of months) and limited plant scale measurement availability which, in addition to the objective, significantly impacted both the high and low level research decisions.

There are four key steps in the dextrose production technology: the fermentation, which is followed by the crystallization, centrifugation and drying. The fermentation of starch impacts significantly the crystallization process: the higher degree of polymerization (DP) oligomers of dextrose acts as growth rate inhibitor, which was confirmed by laboratory scale crystallization experiments. However, the concentration of higher DP components may vary as a function of the fermentation unit operation. Beyond the fine index criteria, the centrifugation is also impacted by the CSD, and, indirectly by the fermentation operation: the smaller crystals increase the cake resistance and also dissolve faster, hence, decreases the overall yield. Although we were able to make minor recommendation with respect to the feeding stream, such as feeding concentration in certain range, adjusting the higher DP concentration and the stream temperature was not possible.

Population balance models (PBMs) are widely applied to describe the crystal size distribution (CSD) variation with time during solution crystallization processes. The dextrose crystallization process in the mixed suspension mixed product removal (MSMPR) crystallizers was described by PBMs involving secondary nucleation and crystal growth, also accounting to growth rate inhibition. The PBMs are solved with the high resolution finite volume method (HR-FVM) in Matlab environment. The kinetic parameters from the literature were used as starting point, and the nucleation and growth rate constants were re-adjusted based on batch laboratory scale measurements [1]. The growth rate equation was extended with a simple empirical growth rate inhibition term, which was estimated using crystallization experiments with pure dextrose solution (no growth rate inhibition) and the plant syrup (known growth rate inhibitor concentration). The performance of the kinetic parameters was evaluated using an uncertain simulation of the current plant operation by a Monte-Carlo sampling of kinetic parameters from the uncertainty space defined by the hyper ellipsoid obtained from parameter estimation. The simulated fine index distribution, after an output error correction, matched the historical fine index data produced by the plant.

The calibrated process model was applied to minimize the fine index taking as decision variables the feeding concentration, temperatures in the crystallizers as well as the recirculation flowrates of each technically feasible configuration. All the optimizations were carried out using 125 randomly sampled kinetic parameters from the known uncertainty space, and the mean values of decision variables were considered as the final solution. The uncertain kinetics based optimization also allowed to compare the various operation scenarios in terms of expected fine index variability: the lower simulated variance of the fine index distribution obtained with the same kinetic parameter combination suggests that the given operation is more robust with respect to the uncertainties and disturbances.

Due to the lack of exact plant-scale model validation, the accuracy of optimizations for the exact fine index prediction might be fair to moderate, however, the results are directly comparable to each other, which allowed to provide a semi-quantitative expected performance of each operation mode but exact comparison of the feasible configurations. The optimization revealed that the existence of recirculation stream has significantly greater positive impact on the fine index minimization than the overall system volume. This was also validated with laboratory scale experiments by a cascade of three MSMPR crystallizers, which clearly showed that the recirculation increased significantly the mean size of crystals in the product stream. This observation enables to decommission the least efficient crystallization units (~40 % of the total volume), leading to significant operation and maintenance cost savings. The optimum feeding concentration, recirculation ratio and temperatures of (operationally) preferred configuration aligned well with the operation of other dextrose crystallization plants of the company which, however, operated at different scales involving different type of crystallizers. Ultimately, the research results gave sufficient support to the company to initiate a 10 year long internal development plan for operating point switch.

References

[1] A. Markande, A. Nezzal, J.J. Fitzpatrick, L. Aerts, Investigation of the crystallization kinetics of dextrose monohydrate using in situ particle size and supersaturation monitoring, Part. Sci. Technol. 27 (2009) 373–388. doi:10.1080/02726350902994050.

573038 Systematic Design and Intensification of Shale Gas Utilization Processes

Recent advancements in hydraulic fracturing and horizontal drilling have led to a boom in shale gas production [1]. The availability of cheap natural gas resources presented many opportunities to the petrochemicals, chemicals, and fuel industries [2]. Because of its wide variety of composition, numerous products can be made from shale gas. Several reforming alternatives of methane are available for producing synthesis gas (syngas) which is a building block for many value-added chemicals [3]. One such chemical compound is methanol, which has varied potential uses. It can be used as a clean-burning liquid fuel; energy carrier; and an intermediate for the production of several vital chemical products like olefins, formaldehyde, methyl tertbutyl ether, dimethyl ether and acetic acid [4-5]. Reforming reactions require high temperatures to operate, which makes these technologies cost intensive [3]. Additionally, as the quality of syngas can dictate the type of end-product and reaction products are needed with high purity, several energy-intensive separation methods are utilized. Systematic design methods can be highly useful, here, to screen, generate and yield novel intensified processing routes. However, traditional conceptual design methods rely on unit operation-based representations which may prohibit identification of such innovative solutions. To this end, recently introduced building block-based representation [6] provides several unique advantages for identification and automatic generation of intensified flowsheet variants.

In this work, building block-based design and intensification approach is used for the synthesis and intensification of shale gas utilization systems. Building block-based superstructure provides an optimization-based methodology toward incorporating process intensification in the conceptual design stage [7]. It is formed by collecting building blocks in a two-dimensional grid and can be used to automatically generate many different flowsheet variants [8]. In this representation, several physical and chemical phenomena, e.g. reaction, vapor-liquid equilibrium, gas-liquid equilibrium, etc., can be represented either as a 'single block' or via multiple neighboring blocks. For example, membrane modules are represented as a collection of two blocks separated by a membrane boundary. Similarly, vapor-liquid contact is represented as two neighboring blocks sharing a common boundary for phase contact. We formulate this block superstructure as a mixed integer nonlinear programming optimization (MINLP) problem. This MINLP model is then solved with cost and sustainability objectives to obtain several intensified flowsheet variants. We will present several intensified flowsheet alternatives that yield less total annual cost compared to the base-case design reported in the literature [9].

References:

[1] Ridha, T., Li, Y., Gençer, E., Siirola, J. J., Miller, J. T., Ribeiro, F. H., Agrawal, R., *Valorization of Shale Gas Condensate to Liquid Hydrocarbons through Catalytic Dehydrogenation and Oligomerization*, Processes 2018, 6, 139.

[2] Joskow, P. L, *Natural Gas: From Shortages to Abundance in the United States*, American Economic Review 2013, 103(3): 338–343.

[3] Medrano-García, J. D., Ruiz-Femenia, R., Caballero, J.A., *Multi-objective optimization of combined synthesis gas reforming technologies*, Journal of CO_2 Utilization 22 (2017) 355–373.

[4] Arora, A., Iyer, S. S., Bajaj, I., Hasan M. M. F., *Optimal Methanol Production via Sorption-Enhanced Reaction Process,* Industrial & Engineering Chemistry Research 2018 57 (42), 14143-14161.

Techno-Economic Assessment and Environmental Impact of Shale Gas Alternatives to Methanol, ACS Sustainable Chemistry & Engineering 2014 2 (10), 2338-2344.

[6] Demirel, S. E., Li, J., and Hasan, M. M. F., *Systematic Process Intensification using Building Blocks*, Computers and Chemical Engineering 2017, 105, 2-38.

[7] Li, J.; Demirel, S.E.; Hasan, M.M.F., *Process Synthesis using Block Superstructure with Automated Flowsheet Generation and Optimization*, AIChE Journal, 2018, Vol. 64, Issue. 8, 3082-3100.

[8] Demirel, S. E., Li, J., and Hasan, M. M. F., *A General Framework for Process Synthesis, Integration, and Intensification,* Industrial & Engineering Chemistry Research (2019), DOI: 10.1021/acs.iecr.8b05961.

[9] Ehlinger, V. M., Gabriel, K. J., Noureldin, M. M. B., El-Halwagi, M. M., *Process Design and Integration of Shale Gas to Methanol,* ACS Sustainable Chemistry & Engineering 2014 2 (1), 30-37

572035 Evaluation of Helium Extraction Technologies While Considering Optimization and Integration with Nitrogen Rejection

Helium is a widely used component in MRI scanners, aerospace and aircraft manufacturing, industrial-leak detection systems, electronics and fiber optics fabrication, welding, nuclear industry and research facilities. It is the second most abundant element in the universe after hydrogen. However, due to its light weight, it easily escapes gravity and is exceedingly scarce in the Earth. Although the Earth's atmosphere is the main terrestrial inventory of helium, it has been widely concluded that extracting helium from the air is not viable due to its very low concentration of 5 ppm. Hence, the only economical helium source is natural gas (NG) reserves with 0.3 to 2 mol% in so-called helium-rich NG fields and as high as 8 mol% in some fields. The USA is so far the world's leading helium supplier with a share of nearly 55% in 2016, followed by Qatar (33%), Algeria (6%), Australia (3%), Russia (2%) and Poland (1%). Helium is marketed in two specifications: crude helium (50-70 mol%) and purified helium (>99.99 mol%).

The extraction of helium from natural gas typically requires four processing steps. The first step is to remove the typical impurities in the gas, namely acid gases, water, and mercury. The second step involves the extraction of heavier hydrocarbons. The third step is nitrogen rejection, which separates a mixture of nitrogen and helium from the product gas. The final step is to recover helium (crude or upgraded [>90 mol%]) from the nitrogen-rich stream, which can be either a cryogenic distillation-based or a membrane-based process. The upgraded helium is then purified to 99.99 mol% helium using pressure swing adsorption (PSA).

Most NG reserves with high helium contents have much nitrogen. This is why both nitrogen rejection and helium extraction are required simultaneously. Thus, there is a strong incentive to integrate them in order to reduce the cost of helium. Several articles (Hamedi et al., 2018; MacKenzie et al., 2002; Rufford et al., 2012) compare the performance of a standalone NRU (Nitrogen Rejection Unit). However, an attempt to examine its integration with a cryogenic distillation-based HeXU (Helium Extraction Unit) is missing in the literature. Alders et al. (2017) compared cryogenic distillation with membrane separation for helium extraction from an NRU. However, they used a hypothetical simulation for NRU to derive the feed to HeXU. In other words, there was no real integration between NRU and HeXU and no optimization.

This study proposes two integrated HeXU and NRU processes, one based on cryogenic distillation and the other based on membrane separation. Then, it optimizes each integrated process using Particle Swarm Optimization (PSO) for minimum power consumption. The optimization is done for various scenarios such as crude helium, upgraded helium, and low versus high pressure nitrogen product. The optimum process solutions are then compared economically to derive a comprehensive map for selecting the right technology for helium extraction. The results show that when the nitrogen is considered effluent and released into the atmosphere, the cryogenic distillation technology outperforms the membrane system due to the significant advantages in both power consumption and capital costs. The cryogenic process can reduce power consumption by 10-40%. In contrast, when the nitrogen product is needed at high pressure and the helium concentration is not very small in the HeXU feed, the MBHeXU consumes less energy compared with the CDBHeXU (maximum 5% reduction). However, it still accounts for higher capital costs compared to cryogenic distillation separation due to a more expensive compression system.

Alders, M., Winterhalder, D., Wessling, M., 2017. Helium recovery using membrane processes. *Separation and Purification Technology, 189*, 433-440. doi:10.1016/j.seppur.2017.07.084

Hamedi, H., Karimi, I. A., Gundersen, T., 2018. Optimal cryogenic processes for nitrogen rejection from natural gas. *Computers and Chemical Engineering, 112*, 101-111. doi:10.1016/j.compchemeng.2018.02.006

MacKenzie, D., Cheta, I., Burns, D., 2002. Removing nitrogen. *Hydrocarbon Engineering, 7*(11), 57-63.

Rufford, T. E., Smart, S., Watson, G. C. Y., Graham, B. F., Boxall, J., Diniz da Costa, J. C., May, E. F., 2012. The removal of CO2 and N2 from natural gas: A review of conventional and emerging process technologies. *Journal of Petroleum Science and Engineering, 94-95*, 123-154. doi:10.1016/j.petrol.2012.06.016

571864 Kinetic Modeling and Process Control for the Preparation of Graphene Oxide

Graphene oxide (GO), as a precursor for graphene synthesis, has received widespread attention due to its potential scalability, excellent dispersibility in various solvents, and facilitates processability toward many practical applications. However, the quantitative description and mechanism investigation for the preparation process of GO still remain a challenge, which essentially hinders the large-scale controlled production of high-quality GO and its related graphene materials. In this work, oxidation process of acidic oxidizing medium with flake graphite was interpreted as a solid-liquid two-dimensional advance of the reaction interface model. Reaction rate constant and kinetics parameters were obtained, and applied to process simulation, optimization and products control. Meanwhile, the exfoliation extent and lateral size distribution of GO during the ultrasonic-assisted exfoliation process were systematically studied for the product layer number and lateral size control. Kapur function analysis was used to arrive at the breakage kinetics under different conditions, and the breakage mechanism of graphite oxide is attributed mainly to sheet fracture. The results can be applied for designing desired industrial process for the preparation of GO with quantitatively controlled product quality by simple process parameters.

558185 High-Fidelity Modeling and Optimization of Natural Gas Dehydration and Liquid Recovery Plants

This study presents the simulation and optimization analysis of a commercial natural gas dehydration and liquids recovery plant based in the region of Abu Dhabi, UAE. gPROMS process builder, an advanced modeling tool based on equation oriented approach, is employed for modelling the gas plant system. In this work, we developed a dynamic model for the gas dehydration system and a steady state model for the natural gas liquids recovery system.

We first show the comprehensive data reconciliation and models' validation performed to ensure that the models' predictions match the real plant behaviour. We then present how the advanced optimization features result in the best operational strategies for the natural gas liquids recovery system. We also show the optimization analysis carried out for maximizing the adsorption cycle time for the dehydration unit while minimizing the total heating duty required for the regeneration of the molecular sieve beds.

The optimization analysis has led to an estimated increase in annual net revenue of natural gas liquid recovery unit by 6 MM USD as a result of maximized liquid hydrocarbon production and lower operating costs related to steam and refrigeration.

567372 Bayesian Optimization Approach to Operation Recipe Optimization of Semi-Continuous Process

Semi-Continuous process is similar to semi-batch process in that reactants are added repeatedly during the operation. However, the main difference is that the product is removed repeatedly in the semi-continuous process, unlike the semi-batch process where the product is removed only once at the end of reaction. The optimization variables for the operation recipe of semi-continuous process include the amount of reactant and the time points to replenish. In the case of the process of which the amount and state of reactant in the reactor is not measurable, the recipe optimization problem is a kind of hyperparameter optimization. Since the time points to replenish are not predictable, the process operates according to the specified operation recipe set before beginning the operation. The optimization also involves dynamic simulation of the semi-continuous process, which requires a long simulation time to ensure that operational constraints are satisfied during the operation.

Not only a long dynamic simulation time is required, but the sequence of replenishing the reactant requires many iteration steps to converge the non-linear solver since the mass and energy change stiffly. In addition, it also takes much time to simulate all the combination of the optimization variables. In this study, a Bayesian Optimization Algorithm [1], which is one of the machine learning algorithms for hyperparameter optimization, is utilized to reduce the number of dynamic simulation runs for recipe optimization. First, a base-case suboptimal recipe is employed to sequentially replenish the reactant at the time when the reactant is depleted. And then new operation recipes are suggested to minimize the amount of reactant used and to maximize the period of replenishment.

The process presented in this study is a pilot-scale carbonation process that captures 40 tons of carbon dioxide per day using 20wt% aqueous solution of calcium hydroxide with a gas flow containing 16 vol% carbon dioxide and consists of two reactors. The carbon dioxide is dissolved in water slurry mixed with a calcium hydroxide, which results in precipitation of calcium carbonate. In each reactor, the reactant and product have a replacement cycle, and gaseous carbon dioxide passes continuously. This dynamic model was constructed using Aspen Custom Modeler (ACM) V10 owing to the limited availability of the physical properties of the involved species. Reaction kinetic parameters were estimated using the data from real operations as well as the literature [2-5].

For the operation recipe optimization, the six hyperparameters such as maximum and minimum liquid level are selected and the range is determined. This set of hyperparameters is optimized by solving the bayesian optimization algorithm that is carried out by linking MATLAB to ACM. First, the next set of hyperparameters is decided by optimizing the acquisition function using the sampling data from the simulation. Then, after the ACM dynamic model has been simulated with the set of hyperparameters, the results are sent to the algorithm to find the next set of hyperparameters. As a result, two new operation recipes are suggested and the optimal sets of hyperparameters for each case are obtained.

[1] Snoek, J., Larochelle, H., & Adams, R. P. (2012). Practical bayesian optimization of machine learning algorithms. In *Advances in neural information processing systems* (pp. 2951-2959).
[2] Cents, A. H. G., D. W. F. Brilman and G. F. Versteeg (2005). "Absorption in carbonate/bicarbonate solutions: The Danckwerts-criterion revisited." Chemical Engineering Science **60**(21): 5830-5835.
[3] Weisenberger, S. and A. Schumpe (1996). "Estimation of gas solubilities in salt solutions at temperatures from 273 K to 363 K." AIChE Journal **42**(1): 298-300.
[4] Johannsen, K. and S. Rademacher (1999). "Modelling the Kinetics of Calcium Hydroxide Dissolution in Water." Acta hydrochimica et hydrobiologica **27**(2): 72-78.
[5] Velts, O., M. Uibu, J. Kallas and R. Kuusik (2011). "Waste oil shale ash as a novel source of calcium for precipitated calcium carbonate: Carbonation mechanism, modeling, and product characterization." Journal of Hazardous Materials **195**: 139-146.

568722 Optimal Scheduling of Multi-Product Pipeline Distribution with Consideration of Comprehensive Transmix Handling for Petroleum Refineries

Pipeline is more efficient and reliable to transfer large amounts of petroleum products than other ways via road, railroad, and vessel. To save capital and operating costs, one pipeline needs to transfer different types of oil products to different oil depots according to their ordinary demands. As two consecutive different types of oil products inside a pipeline will not be efficiently separated by a physical divider, the transmix (TM) will be generated in the interface between oil products. To guarantee the oil quality of product adopted by storage tanks at depots, a considerable amount of TM products along the pipeline needs to be handled properly. In reality, TMs can be handled by blending, distillation, or downgrading operations. Here, the downgrading operation means a TM mixed with similar types of oil products with different grades (e.g., premium and regular gasolines), could be considered as the lower-grade one (e.g., downgrade TMs mixed with premium and regular gasolines to the regular gasoline). The costs for blending and distillation comes from the storage, blending operation, transportation and reprocessing, while the cost for downgrading is a loss of margin which comes from the different price between the two mixed refined products. Thus, the objective function aiming at minimizing the total cost may ignore the potential benefit contributed by downgrading operations. Therefore, it is worth the effort to set the scheduling problem objective to maximize the total operating profit along the scheduling time horizon by optimizing the injection and transportation sequence as well as the transfer amounts of different oil products.

In this work, a new continuous-time based MILP model under the goal of maximizing the total operating profit is developed to optimize the distribution and inventory management of multiple oil products for petroleum refineries. The studied scope includes multiple refinery oil products, a long-distance pipeline, and several depots which consists of various storage tanks and transmix tanks dedicated to hold different pure products and TMs, respectively. The TM generated inside the pipeline will be handled with a combined method including blending, distillation, and downgrading operations. The TMs with different types of oil products (e.g., gasoline and diesel) will be handled by blending product-rich mixtures into pure products while recycling and distilling well-mixed TM portions. Meanwhile, the TMs mixed with similar types of oil products with different grades will be handled by downgrading operations. Overall, the scheduling model could provide optimal solutions including injection sequence and volume of oil product into pipeline, oil slugs movement inside pipeline, discharging product to multiple depot tanks, as well as exporting oil product based on the demands of consumer markets. The efficacy of the developed scheduling model has been demonstrated by multiple case studies.

Keywords: Multi-product, Scheduling, Pipeline, Transmix (TM), Blending, Distillation,

Downgrading, MILP, Maximizing profit

581026 Digital Twin, Raising It out of the Realm of a Cliché

The Engineering and Industrial world are lagging manufacturing industries to truly embrace digitalization. AVEVA is accelerating the digital transformation journey of their customers and disrupting their current business processes to add value opportunities to their businesses. A common disruptive technology is the concept of Digital Twin. Through our work with our customers, we realize that believe to define a Digital Twin for the industrials we should draw a parallel to a the actual Human Twin. A twin has a shape, behavior, senses and needs to work for 100 years. The same is true for the digital twin of an industrial asset whether that is a an offshore platform, a refinery, a power generator or a ship. The Digital Twin of a plant is defined by engineering data (digital shape), engineering models (digital behavior), real-time series data (digital senses) and finally maintenance data (digital health). Once this data is connected just like a human it is can be stored in memory (historian), analyzed to make the right decisions (predictive analytics), learns from its mistakes (machine learning), asks questions (cognitive thinking) to make it more efficient (APC, real-time optimization), be mobile (smart devices) and finally visualize data in context (3D, X Reality etc...). AVEVA will present how it is connecting engineering, maintenance and operations data, people and business processes to bring the digital twin out of the realm of a Cliché.

580756 Asset Optimization Software Suites to Drive Digitalization – from Process Modeling to Optimization to Asset Optimization

Process Simulation has undergone significant transformation over the past decades driven by innovations in engineering best practices catalyzed by chemical engineering, mathematical optimization and software innovations. It has come a long way from automating the design calculations and process modeling to an environment for process optimization accelerating collaboration across the disciplines and tackling optimization across multiple dimensions simultaneously. The use of Process Simulation has also expanded from the traditional design applications to operations and planning. This digital transformation has created significant value in the industry resulting in capital and energy savings combined with better yields and capacities, increased safety and dramatic improvements in engineering productivity and production. Rapid advances in machine learning, advanced multivariate analytics, high performance mobile computing, and equipment and sensor connectivity are playing a role in enabling a whole new generation of technological solutions. The paper will cover the vision for Process Simulation including expansion from Process to Asset Optimization with first principles-based models and data-based machine learning models for process engineering, operations and performance management.

580855 A General Digital Applications Platform

Deep process knowledge captured within physics-based models is a key element for the successful digitalization of process industries. Digital applications making use of models derived from first principles are increasingly being used in industrial practice, particularly in the area of process operations. They include both open-loop applications, such as model-based soft sensing and monitoring of equipment degradation, and closed-loop ones such as Real-Time Optimization and nonlinear Model Predictive Control.

However, most non-trivial digital applications involve much more than a mathematical model being solved in a simulation or optimization mode. They often require multiple model-based calculations being scheduled over time and exchanging data with each other, while potentially being subject to occasional failures. They also involve extensive communication with external data servers, such as distributed control systems, plant historians, commercial databases and user dashboards. Moreover, the data involved in all such communications may be subject to systematic and/or random errors, and may even occasionally become unavailable.

This paper describes a recently developed general software platform for resilient and sustainable digital applications, taking account of the above considerations. The platform significantly reduces the cost and increases the reliability of development, testing, deployment and maintenance of diverse applications within a unified software architecture. Industrial examples illustrating the flexibility of the proposed design are presented.

580649 Concept of Digital Twin and Digital Thread in the Chemical & Process Industry

The vision of digital twins involves the digital twin of the product, the digital twin of process performance and the digital twin of operation.

In the R &D, stages product development and recipe transformation in with digital models can accelerate innovation. In design & engineering stages, all the disciplines involved in a plant project can generate a shared data model –saving valuable time and expenses and at the same time increasing engineering quality. The digital twin also allows the plant to be simulated prior to the critical commissioning phase, which ensures a reliable implementation. The operators can also receive targeted training before the system is commissioned. The digital twin of the operation significantly increases the reliability and availability of your plant by continuously monitoring your mechanical assets using IIoT.

The realization of this needs cross-organization collaboration. Realization of the concept needs various tools, software, hardware and collaboration platforms to work together. This talk with explore those possibilities and readiness.

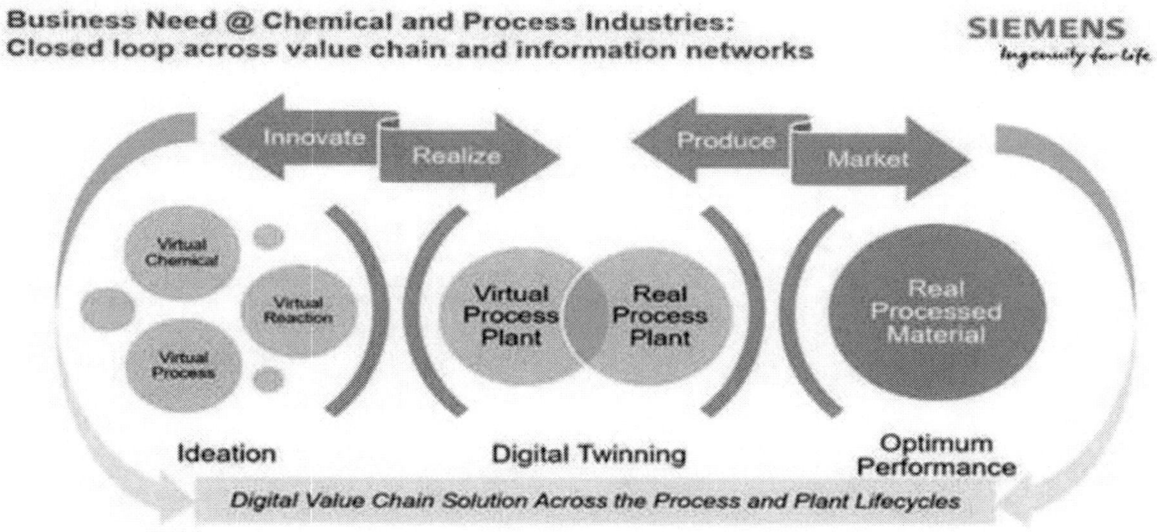

571319 Digital Twin for Aluminum Smelting

Building digital twins for manufacturing processes like aluminum smelting using first principles modeling has been considered challenging owing to the scale and complexity of such processes. A considerable mismatch is normally observed between measured data and predicted values of desired variables. This mismatch is usually because of uncertainty in the information fed to the model such as model assumptions, parameters, correlations from literature, etc. A systematic approach to identify such missing information would be to pose the model as an input-reconstruction problem. While a typical input reconstruction problem is solved using optimization-based algorithms where the objective would be to minimize the deviation between data and prediction, it may not be suitable for complex processes such as aluminum smelting. In this work we present a novel method to reconstruct missing information using a combination of first principles modeling and data analysis. The input reconstruction problem is reformulated as a set of differential algebraic equations (DAE) using measured data. Solving these DAEs help identify the missing input information and other unknown states in one shot. This method gives a complete data set (reconstructed inputs and measured data) which is used to train machine learning (ML) algorithms to build new correlations and predict unknown inputs directly in the digital twin model thereby giving more accurate predictions of desired process parameters.

579640 Transforming Abundant Data into Improved Process Reliability (Keynote Talk)

Today, processing facilities are beginning to effectively utilize the abundance of process and asset health data and information as part of an overall IIoT strategy. Although a digital transformation will require investments in technology, one cannot forget the investment in training staff to behave differently (improved) with the new timely information. It should be noted that some jobs will be replaced (e.g. manual data collection), new roles will emerge (e.g. analysts and planners), and other jobs will evolve to be more efficient. Properly trained operations are invaluable in this process, as they not only act on information from analyzed data, but can also make decisions on their own based on sensor data.

Abnormal operation and imminent asset failure can quickly be identified and corrected with properly trained staff to act on this new timely analyzed information. Big data and IIoT will continue to have a role with shortages of skilled resources, thus having the staff be more efficient and effective at their role with additional information and insight into operations.

This presentation will focus on the applications where big data and IIoT have improved process and asset reliability. Although the applications presented will come from oil refineries, the solutions can be applied to other continuous processing facilities.

579639 Going Digital in Downstream; Refinery of the Future (Keynote Talk)

The last US refinery was built in 1977, to bring these refineries into digital capability is a significant effort and we will explore actual examples and the value that refiners have received. OSIsoft participates in a project to help define the Refinery of the Future. Several examples, like a sensored pump, best uses of IIOT, 3D visualization of the plant and integrity of the asset will be shared to highlight how the downstream business is addressing these technological opportunities. We will share best practices and the progress of recognized leaders in the downstream digital transformation.

579317 Emission Monitoring and Plant Integrity Enhancement

Historically, refineries and chemical plants contribute significantly to emissions of Volatile Organic Compounds (VOCs) and Hazardous Air Pollutants (HAPs). With the increasingly stringent regulations and laws, emission management has to go beyond the traditional Continuous Emissions Monitoring System (CEMS), Leak Detection and Repair (LDAR) approach to prevent potential emission events.

To comply with the new requirement, one prescriptive approach is to enhance plant integrity. Alternative monitoring technologies, such as wireless sensing, optical sensing, facilities the engineers and operators to identify the leaks/emissions conveniently. Information technology, such as industrial 4.0, data analysis and Industrial Internet of Things (IIoT) enables the process industry to communicate and interact effectively. By integrating these technologies, plant integrity, such as predictive analytics, predictive maintenance can be conducted to improve Health, Safety, and Environment (HSE) of process industry.

This presentation describes the relevant environment regulations, emission monitoring practices, and state-of-art emission management technologies, such as predictive flare control and predictive maintenance. Cases of deployments will also be discussed in this presentation.

565604 Optimizing Real-Time Spatiotemporal Sensor Placement for Monitoring Air Pollutant for Health Impact Assessment

Air pollution exposure assessment, which involves the monitoring of pollutant species concentrations in the atmosphere, is an important step in assessing the health impact of air pollution. Currently, the air pollutants are often monitored via stationary monitoring stations. Due to the high cost of maintaining a monitoring network, the monitors can only be installed in limited number of locations. The sparse spatial coverage of monitors can lead to errors in estimating the actual exposure of pollutants in individuals. The recent advancement of cheaper sensors can help address the limitations of using stationary monitors. One approach is dynamic sensing – a new monitoring approach that adjusts the locations of portable sensors in real time to measure the dynamic changes in air quality. The key challenge in dynamic sensing is to develop algorithms to identify the optimal sensor locations in real time in the face of inherent uncertainties in the fate and transport of the air pollutants. The objective of this paper is to develop an algorithmic framework to address the challenge of sensor placement in real time. We demonstrate the capability of our algorithmic framework in a case study in Atlanta. Our real-time sensor placement optimization algorithm will allow, for the first time, the assessment of spatial-temporal variability of pollution which cannot be accomplished by a stationary monitoring station. Besides air pollution exposure assessment, our algorithm can have broader applications in improving sensor placement in drone-based monitoring for water bodies, water security networks, or advanced power systems.

562063 Energy Harvesting Sensors Using Magnetic Phase Transition

Many countries have recently become interested in cyber-physical systems (CPS) in industry and society for sustainable development. CPS equips a data acquisition function, a data storage function, and a network to transfer data. To deploy CPS, it necessary to develop overall security systems, to increase their energy efficiency and to find new energy sources. To overcome these energy-related issues, in this research we investigated the possibility of integrating data acquisition sensors with a recently developed energy harvesting system that combines the magnetic phase transition resulting from changes in temperature, and electromagnetic induction. The proposed system can provide wireless temperature or velocity sensors that directly measure electromotive forces generated by Faraday's law of induction without any additional energy input. Our proposed energy harvesting sensors have the potential to contribute significantly to the development of CPS in the near future.

565916 Bio-Based Benzoxazine Monomers and Polymers Based on Difuran Diamine(Dfda)

A new class of benzoxazine monomers based on difuran diamine (DFDA) obtained by the coupling of furfuryl amine using formaldehyde will be discussed. The benzoxazine monomers were synthesized by reacting DFDA with a number of bio-based phenolic compounds and formaldehyde. These systems were purified by precipitation. Depending on composition, the resulting solid powders melt at temperatures ranging between 70°C and 150°C to form low viscosity liquids that can be readily used to impregnate fiber reinforcements. Onset cure temperature also varies depending on composition in the range of 120°C-240°C. The resulting polybenzoxazines have Tg's ranging from 220°C to 320°C. It was found that adding furan rings into the backbone of the polybenzoxazine network provides the added benefit of 800°C char yield as high as 70% as measured by TGA temperature ramps in an inert environment. In addition to promising processing and thermal characteristics useful for applications requiring high temperature ablation resistance, these materials contain high renewable content.

556622 Overcoming the Influence of HMF Reaction Mixture Constituents on Extraction Solvent Performance

HMF (5-hydroxymethylfurfural) is a promising bio-based platform chemical that is obtained from hexose dehydration and can be used to produce biofuels and various chemical products [1]. The hexose dehydration reaction is equilibrium limited, reversible and the HMF formed in this reaction can decompose into levulinic acid (LA) and formic acid (FA). In order to avoid HMF degradation, extensive research has been performed on biphasic systems wherein the catalytic dehydration reaction is coupled with direct extraction of the HMF into an organic solvent to increase the selectivity as well as conversion [2].

The focus of this work is a systematic study how the constituents of the reaction mixture affect the extraction performance of organic solvents. A solvent screening has shown that in addition to the widely studied solvent methyl isobutyl ketone (MIBK), 2-pentanol is a promising alternative with comparable selectivity but significantly higher distribution coefficient. Next to water and HMF the reaction mixtures typically contain the hexose substrate (glucose, fructose), salts, catalyst (acid) and co-solvents (DMSO, ionic liquids, deep eutectic solvents) [1]. All these constituents will influence the extraction performance (distribution coefficient, selectivity) of the solvent. However, while using these complex reaction mixtures in studying the effect on conversion and selectivity of the dehydration reaction, there are to our knowledge no studies that have systematically investigated the combined effect of these constituents on the solvent performance.

The study has been executed by determining liquid-liquid phase equilibrium data for the ternary systems HMF, water, and solvent (MIBK or 2-pentanol) at 313.15 K (40oC) and atmospheric pressure. Inorganic salts (NaCl, KCl, Na2SO4, KCl, K2SO4), [EMIM][BF4] as ionic liquid, 1:2 molar ratio mixture of choline chloride/urea as deep eutectic solvent and fructose were added to the aqueous phase. All data obtained were correlated using the NRTL model. The effects on solvent performance (HMF distribution coefficient and the selectivity) were derived from the LLE data.

The results indicated that higher distribution coefficients are achieved in the 2-pentanol-HMF-water system compared to the MIBK-HMF-water system. On the other hand, the separation ability of MIBK is better than that of 2-pentanol. Adding up to 10 wt% inorganic salts to the system enhanced the distribution coefficient of HMF into the organic phase as well as the selectivity up to 2 times confirming the salting-out ability of the studied salts. The addition ionic liquid showed the opposite effect. At 30 wt% [EMIM][BF4] the HMF distribution coefficient as well as selectivity were reduced to 1/3 of the original value. This salting in effect is caused by the specific interactions between the HMF and ionic liquid that is known to increase the reaction selectivity by complexing with the hydroxyl groups of HMF. For the deep eutectic solvent a comparable effect as with the ionic liquid was observed. Also the introduction of fructose in the system reduced the HMF distribution coefficient and selectivity. Typically substrate concentrations up to 30 wt% are applied and at these conditions a significant decrease in solvent performance is observed. Finally it was observed that the presence of HMF and fructose in the system reduced the solubility of especially the divalent inorganic salts.

Overall monovalent salts like NaCl and KCl are the most suitable candidates to overcome the negative effects of co-solvents (ionic liquid, deep eutectic solvent) and hexose substrate on the solvent performance.

[1] Van Putten, R. -J.; Van der Waal, J. C.; De Jong, E.; Rasrendra, C. B.; Heeres, H. J.; De Vries, J. G., Chem. Rev. 2013, 113, 1499-1597
[2] E.C. Sindermann, A. Holbach, A.B. de Haan, N. Kockmann, Chem. Eng. J. 2016, 283, 251-259A. Bianchi, N.C. Jones, Chem. Eng. J. 157 (2019) 326–337.

571711 The Chemical Properties Allowing the Tetrabutylphosphonium Chloride-Water Mixture to Efficiently Dissolve Cellulose in High Water Concentrations Via Molecular Dynamics

Our society is gradually moving away from fossil-based fuels, as we develop and transition to renewable fuel alternatives. Many of these renewable energy sources such as wind, solar, and tidal power can easily distribute power to stationary locations, via our existing electric grid. However, we also need to provide power to mobile sources, such as cars, tractor trailers, cargo ships, and other heavy equipment. In the near future, renewable biofuels can supply the energy dense and portable power to our vehicles, heavy equipment, and the shipping industry, much like gasoline and diesel do today.

These biofuels can be derived from waste biomass, such as corn husks at the end of a growing season, waste tree limbs from logging operations, or sawdust from a sawmill. The cellulose, contained in the biomass, needs to be broken down before the enzymes or other catalysts can turn it into fuels, such as ethanol. Many processes currently require high temperatures to sufficiently break down the cellulose, so it can efficiently be catalyzed in the downstream processes. Since a large part of this energy expenditure is due to the elevated temperature, it would be economically ideal to operate at near room temperature. It's also critical that these solvents work in the presence of water, since cellulose naturally contains water, and an additional pre-drying step may thwart our efforts in developing an economical and sustainable process.

There are a variety of ionic liquids (ILs) capable of dissolving cellulose. However, many ILs fail to dissolve cellulose when even small amounts of water are added to them at moderate temperatures.[1-5] The IL and co-solvent mixtures of tetrabutylphosphonium hydroxide (TBPH) and water are capable of dissolving at least 15 wt% of cellulose at room temperature, within a working concentration range of 30 to 50 wt % water.[6] Therefore, studying the less reactive derivatives of the TBP class of ILs, such as tetrabutylphosphonium chloride (TBPCl), could lead to an optimized IL and co-solvent mixture for dissolving cellulose.

In this work, we analyze the TBPCl-water mixtures physical and chemical interactions that take place during the cellulose dissolution process. We investigate the hydrogen bonding strengths and disruption capabilities between the TBPCl-water mixture and the cellulose bundle. The chloride anion and TBP cation show similar binding and cleaving properties as previous studies.[7] Water plays a supporting role by assisting in the cellulose separation process and preventing cellulose reformation, providing the TBP molecule with added time to attack and cleave the weaken cellulose region. The pairwise interaction energies of the system are also calculated, which portray favorable interactions between the cellulose and the IL-water mixture. Understanding the physical and chemical properties that allow the TBPCl-water solution to work efficiently in water, is critical to developing more economical IL-water mixtures in the future.

[1]Fang, Z., Smith, R. L., Jr., Qi, Z., Eds. Production of Biofuels and Chemicals with Ionic Liquids; Springer: Heidelberg, 2014.

[2]Wang, H.; Gurau, G.; Rogers, R. "Ionic Liquid Processing of Cellulose". Chem. Soc. Rev. 2012, 41, 1519–1537.

[3]Fang, Z., Smith, R. L., Jr., Qi, Z., Eds. Production of Biofuels and Chemicals with Ionic Liquids; Springer: Heidelberg, 2014.

[4]Mazza, M.; Catana, D.-A.; Vaca-Garcia, C.; Cecutti, C. Influence of Water on the Dissolution of Cellulose in Selected Ionic Liquids. Cellulose 2009, 16, 207–215.

[5]Swatloski, R. P.; Spear, S. J.; Holbrey, J. D.; Rogers, R. D. Dissolution of Cellulose with Ionic Liquids. J. Am. Chem. Soc. 2002, 91, 4974–4975.

[6]Mitsuru Abe, Yukinobu Fukaya and Hiroyuki Ohno, 2011, Fast and Facile Dissolution of Cellulose with Tetrabutylphosphonium Hydroxide Containing 40 wt% Water

[7]Brooks D. Rabideau, Animesh Agarwal, and Ahmed E. Ismail, 2013, Observed Mechanism for the Breakup of Small Bundles of Cellulose 1α and 1β Ionic Liquids from Molecular Dynamic Simulations

558260 Martini Coarse-Grained Model for Poly-e-Caprolactone in Acetone-Water Mixtures

In the last decades, biodegradable polymer nanoparticles (NP) gained a lot of attention in the chemical engineering industry, playing a key role in many fields, such as cosmetics, pharmaceuticals and textile industries. Besides, in the last years a strong focus has moved on green and sustainable nanotechnology, in which nanoparticles are employed as the pillar of a new sustainable industrial world. This spans from agriculture to water and clean energy applications.

Due to the complexity of the phenomena involved in the nanotechnology processes, the solely experiments turn out to be not sufficient for a complete description thereof. A multiscale modelling approach is therefore necessary. Here, we present the interesting case of polymer NP produced via solvent displacement in a process called flash nano-precipitation (FNP), for poly-e-caprolactone (PCL) in acetone/water mixtures. The main application of such a process is the NP production for drug-delivery systems, but it is worthwhile to mention that this modelling approach has a general validity and can be transferred to other NP formation processes.

In this multiscale approach, three different scales are investigated and interconnected between themselves: the molecular scale, the NP/clusters scale and the mixer (or continuum) scale. The NP/clusters dynamics is modelled by a population balance model (PBM), built upon MD simulations, and coupled, in turn, with computational fluid dynamics, in order to predict the kinetics effects on NP formation. Here, particular focus is given to the molecular scale, in which the given system is studied by means of both all-atom (AA) and coarse-grained molecular dynamics (CG-MD).

AA-MD is employed to overcome the very well-known issue of acetone/water de-mixing in MD simulations, despite these two solvents are completely miscible at room temperature and any proportion. A charge-on-particle model for acetone is introduced, in order to reproduce the correct polarization response in water-rich environments. The basic dynamical properties, e.g. viscosity and diffusion coefficients, are shown to have a better agreement with experiments, compared to the classical atomistic force fields.

CG-MD is used to build up a molecular model for the PCL in solution, parametrising the non-bonded potentials by means of a hybrid thermodynamic-structural approach. The CG model is based on the MARTINI force field and is shown to correctly predict the mean radius of gyration for short and long single polymer chains in mixture showing a good agreement with the Flory's law for real polymers in both good and bad solvents. The models presented here have a general validity and pave the way to future studies of the self-assembly of multiple chains in the simulation box, together with the investigation of different solvents effects, as well as the presence of the drug/active principle, still open topics in many industrial fields.

561276 Lignin-Based Functional Polymers

Lignin is an important source of synthetic materials because of its abundance in nature, low cost, stable supply, and no competition to the human food supply. The crosslinked phenolic polymer, lignin, contains a large number of aromatic groups that can be used as a substitute for petroleum-based aromatic fine chemicals. However, modification of lignin is necessary for its application in advanced materials due to its chemically inert nature and structural complexity. Polymeric modification of lignin via graft copolymerization represents an important avenue for modification because this method forms stable covalent bond linkages between lignin and synthetic functional polymers. Through the lignin graft copolymerization, lignin was integrated with synthetic polymers to obtain certain smart functions. In this presentation, we summarized our recent founding in thiol-ene lignin modification, and a lignin-based shape-memory polymer, synthesize and special properties.

574488 Uncertainty and Sensitivity Analysis Assessing the Robustness of Models Employed in Superstructure Optimization for the Downstream Process Design of a Biorefinery

An integrated biorefinery is an essential concept for the sustainable production of food, chemicals and power generation, utilizing e.g. lignocellulosic waste from agriculture as feedstock [1]. The crux of this concept is a poor economic potential due to the multitude of process steps for the pretreatment of the feedstock and the downstream processing (DSP) [2]. In terms of the latter, the separation and purification of a desired substance from fermentation broth is a complex and versatile procedure [3]. Hence, classic heuristic approaches in process design for biorefineries will succumb to a methodic approach. Superstructure optimization is a well-proven approach to tackle this exact challenge [4]. In order to utilize it, models for every considerable unit operation have to be developed. Apart from the issue of which unit operations to employ, the predominant question targets the robustness of the implemented models, which exceedingly affect the results of the superstructure optimization.

The considered case is the DSP of xylitol produced by fermentation. Xylitol is a sugar substitute with manifold beneficial health properties and especially suitable for diabetics [5]. Xylitol is highly soluble in water, which is the bulk substance in fermentation broth. Moreover, it contains salts, byproducts and – in the case of lignocellulosic feedstock – inhibitory compounds. Hence, the downstream process has to segregate xylitol from water and the other named compounds. To this end, the unit operations, which are represented in the superstructure by first-order models, are an evaporation unit, a liquid-liquid extraction unit and a crystallization unit. All three are incorporated as unit operation in the separation section of the DSP, where the latter one is also considered as the unit operation for the purification step.

In order to address the arisen question regarding the robustness, a Sensitivity and Uncertainty analysis with the three unit operation models is performed. The Sensitivity analysis is conducted globally and comprises Sobol indices, the Uncertainty analysis is performed by utilizing Monte-Carlo techniques. This key analysis step now provides various insights into the functionality of each of the models and allows for conclusions regarding the following interactions of the model by performing the superstructure optimization. Firstly, the sensitivity analysis serves as design space exploration, yielding information about whether similar parameters in the models are more sensitive. This allows conclusions towards which of them should be included in the objective function of the following superstructure optimization, which is a valuable contribution in order to improve its efficiency. Secondly, the principal question about the robustness is answered by the combined Uncertainty and Sensitivity analysis: with the knowledge about sensitive model parameters and the error-propagating properties of the model valuable insights about the reliability of the interconnection of several of these models are provided. In case the downstream process should involve several repetitive separation and purification steps in order to match the presumed purifications, the overall resilience of the suggested design can be assessed.

With this cornerstone, the following optimization steps for the superstructure and operations optimization of the biorefinery and an extended supply chain analysis based on robust optimization can be performed soundly. Finally, this will contribute to the development of a consolidated, economically viable biorefinery process.

References

[1] F. Cherubini, Energy Conversion and Management, Vol. 51, 2010, pp. 1412-1421.

[2] A. Chandel et al., Bioresource Technology, Vol. 264, 2018, pp. 370-381.

[3] M. Corbetta et al. in: G. Kopanos et al. (Eds.), Advances in Energy Systems Engineering, Springer, Cham, 2017, pp. 683-712.

[4] C. Gargalo et al. in: J. Klemes (Ed.), Assessing and Measuring Environmental Impact and

Sustainability, Elsevier, Oxford, 2015, pp. 277-321.

[5] S. Mussatto in: S. da Silva, A. Chandel (Eds.), D-Xylitol. Springer, Berlin, Heidelberg, 2012, pp. 209-323.

564641 Techno-Economic Analysis for Rapid Commercialization of Chitin Nanofiber Suspensions for Barrier Applications

A significant challenge in the development of sustainable chemicals and plastics is establishing an explicit link between fundamental materials discovery and cost-effective process development that supports commercialization. Techno economic analysis (TEA) is a very valuable tool in the synthesis and design of processes to make biofuels and bioproducts. TEA is at the core of optimized process economics, minimized environmental impact, and establishing the competitive advantage of innovative products over the existing slate. It can be especially useful to identify research and development opportunities that enhance the viability of bioproducts and bring them to commercialization. Chitin is one of the most abundant biopolymers found in crustacean shells, insect exoskeletons and fungi. It's natural production rivals that of cellulose. In its nanofiber form, chitin is found to have excellent oxygen barrier properties comparable with poly(ethylene terephthalate). Chitin nanofibers (ChNF) are produced from purified chitin using high pressure homogenization often requiring multiple passes. TEA was performed on the entire production process of ChNFs from crab shells. The TEA concluded that homogenization is the most energy and capital intensive unit operation contributing to 57 % of the operating cost and 58 % of the installed cost of the manufacturing process. In order to reduce the cost of ChNFs, and increase the ease of homogenization, a controlled deacetylation process was applied to the chitin. The influence of pressure on homogenization process of deacetylated chitin was also studied. It is found that suspensions of deacetylated chitin reach higher light transmission values at $1/3^{rd}$ the number of passes as required by their non-deacetylated form. The nanofibers from this deacetylated chitin were then cast into films and compared with those from non-deacetylated chitin. It was found that controlled deacetylation can increase tensile strength by 150 % and strain at break also was found to increase by 200 %. The oxygen permeability of the deacetylated material was similar to that of the non-deacetylated material. In addition, it was also found that the deacetylated material could be homogenized at lower pressures of around 8000 psi as compared to non-deacetylated material which requires pressures of 22000 psi. TEA on a modified process was performed to discover that the modified process with a deacetylation step reduces the installed equipment cost by 60 %. The resulting manufacturing costs of ChNFs are decreased by 38 %. Additional process modifications were also evaluated, the most significant of which envisioned integration with a pulp mill and the final integrated and streamlined process resulted in a 83 % reduction in cost of manufacture of ChNFs. These developments show how traditional TEA techniques can benefit sustainable materials processing and bring them closer to commercialization.

577532 Best Practices for Teaching Focused Faculty

Between 2006 and 2015 the number of full time teaching focused faculty in Engineering and Technology programs has increased by more than 80 %, yet during approximately the same period the number of full time Tenure/Tenure track faculty grew by about 10 %. Statistics will be provided for the number of chemical engineering faculty in the US by rank, position and gender. An overview of the contributions that can be made by Teaching focused faculty will also be presented. Based on the literature and experiences of Teaching focused faculty, best practices are recommended for the successful, mutually beneficial practices for Teaching focused engineering faculty and their administrative leadership.

580825 Cultivating Support and Success As a Department's First Permanent Non-Tenure-Track Faculty Member

In the fall of 2015, I was hired as the first permanent non-tenure-track faculty member in the Department of Chemical & Biomolecular Engineering at the University of Delaware. The position has brought many challenges but also many opportunities, and I was able to build on my past experiences as a visiting teaching assistant professor and non-tenure track lecturer at two other institutions to help define my position. Over the past four years, I have established roles in my department that make my job fulfilling and full of opportunities for growth.

I believe the University of Delaware is one of the "gold standards" in terms of logistics related to non-tenure-track faculty. Our treatment of this nontraditional track is evident even in the name: we are called Continuing Track (CT) faculty, to define us as something that we are (continuing from contract to often-lengthening contract), rather than something we are not (i.e., on the tenure track). CT professors share titles with tenured/tenure-track (T/TT) faculty, as well as a parallel promotion path; they are afforded sabbaticals every six years; they participate in faculty governance. Our status is further secured through the support of the American Association of University Professors, as well as a CT Caucus that works to improve the quality of life for all CT faculty at the university.

Being the first CT faculty member of a department came with its challenges: the right balance of teaching, scholarship, and service needed to be agreed upon, and the department faculty had to update our by-laws to align with the policies for CT faculty at the university and college level. I have relied on mentorship from CT faculty in adjacent departments (particularly chemistry and mechanical engineering) as well as T/TT faculty in my own department, as no one person has wholly worked under the same expectations as I have. My department chairs have worked carefully with me to ensure that I am given a workload that allows me to thrive as an educator. In this talk I will share elements of the creation and cultivation of my position as an assistant (and now associate) professor on the Continuing Track in chemical engineering at the University of Delaware that I believe to have impacted my ability to excel in ways that are different from, yet parallel to, our T/TT faculty.

558658 Lessons from an Untenured Administrator…

As the number of non-tenure track faculty members grows across the country in engineering disciplines, it is perhaps inevitable that some of us will end up in administrative roles that in the past may have been primarily reserved for those on the tenure track. Department head, assistant or associate department head, director of undergraduate studies, and even associate dean positions are now sometimes being filled by faculty colleagues that are members of a class of faculty known by many titles: instructional faculty, teaching faculty, in-residence faculty, teaching-track faculty, and the like. The names may be different, but the job descriptions are generally the same: we don't do "technical" research in our primary discipline (though we increasingly are getting involved in educational scholarship!), and we do a much larger share of teaching (and potentially service) than our counterparts on the tenure track.

In this presentation, I hope to present some of my observations (and best practices) that I've seen as a tenure track faculty member (2003-2008) as a non-tenure-track faculty member (2009-Present), and as an academic administrator: associate department head (2011-2013) and Associate Dean (2013-Present). NTT faculty can encounter a broad range of experiences in their departments based on both the micro-scale (departmental) and macro-scale (university-wide) culture, and this can lead to both some fantastic opportunities as well as feelings of isolation and overwork. Taking advantage of certain opportunities without the protection or stature that tenure grants can be daunting, but also incredibly liberating. It can also be frustrating when certain advancement paths seem closed to us because the position "requires" a tenured faculty member (for no reason that anyone can seem to clearly articulate). In academia, tenure is an obvious marker of quality and success, and one that has been collectively defined. For those of us not on that path, recognition and advancement can be more nebulous, and often more defined by our personal relationships than by anything else. I hope to share with you some of the ways that I have been successful in navigating the NTT world, and that some of those lessons are transferable to your circumstances and institutions.

580826 Presenting Yourself to the World: Academic Portfolio As a Professional Brand

Do you see yourself as a brand or just a busy academic focused on surviving your teaching, service, and research assignments? You may be working hard, but at the end of the day, a professional brand is the sum total of you and how you'd like to present yourself to the world – or at least specific target audiences (colleagues, students, employers, funders, promotion committees, etc.) The processes of sharing who you are can sometimes be overwhelming and exhausting. However, what if we create leverage our academic expertise and prepare a portfolio that can help us improve our academic career, showcase our work, and build a community in engineering education?

In this presentation you will be able to identify best practices on constructing an academic portfolio that can potentially become part of your professional brand. Also, your will learn how to share it online using multiple platforms. You are encouraged to bring a computer or any material that you want to share.

578186 Panel Discussion: Getting Directions on the Path Less Traveled

This segment of the session will be a panel discussion exploring the journey of non-tenure-track faculty from a variety of perspectives.

576960 A Novel Multiscale Model Approach to Describe Products Formation during the Cultivation of *Haematococcus Pluvialis* Under Different Environmental Conditions

Bioprocess development represents an onward challenge for world sustainability, and economy. Microalgae as bio-factories of the future are currently facing an increasing interest" as they possess a range of interesting properties. The photoautotrophic growth of microalgae is based on the absorption of carbon dioxide as main carbon source, making them a promising tool for dissipating the continuously increasing CO_2 in the atmosphere (Singh & Ahluwalia, 2013). The cultivation of microalgae could be inserted in a wider context of bio-refinery exploitation due to their capability to produce high-value molecules, biofuels, fertilizers, antioxidants, anti-inflammatory, and antimicrobials making them potentially exploitable in a refinery context (Chew et al., 2017). Furthermore, these properties are key aspects to be improved when the bio-refinery concept have to be developed (García Prieto, Ramos, Estrada, Villar, & Diaz, 2017).

Haematococcus pluvialis is able to produce significant amounts of compounds belonging to lipids and starch classes, and various specific pigments such as astaxanthin, and lutein (Shah, Liang, Cheng, & Daroch, 2016). Their production could vary depending on the environmental stress conditions such as light intensity and nutrients concentration (D'Alessandro & Antoniosi Filho, 2016). However, stress conditions lead to undesired phenomena such as cell lysis (Hata, Ogbonna, Hasegawa, Taroda, & Tanaka, 2001), which is likely to be related with products release in the external cell environment and eventually with their loss. Photoautotrophic processes are usually characterised by long term cultivations, which make their exploitation even more difficult when compared with other kind of cultivations such as mixotrophic or heterotrophic. Nevertheless, photoautotrophic growth has been indicated such a convenient option to produce specific compounds as lutein, due probably to the low complexity and economy of the cultivation system (Lin, Lee, & Chang, 2015).

This work aims to develop a novel multiscale model of the photoautotrophic growth to optimise the production of lipids and pigments. *H.pluvialis* undergoes a transition from a vegetative (green) to a cyst (red) stage during the nutrient depletion. Cells in the vegetative stage are dimensionally smaller than cells in the cyst stage, and their membranes are thinner. Hence, cells are susceptible in different ways to cell rupture during nutrient depletion, which leads to cell number loss. Cell growth, product formation, the cell lysis, are described through the development of a segregated model based on Population Balance Equations (PBEs), making possible the description of the relationship between cell dimension/transition, cell loss, and products availability. Cell volume is a key state of the population balance model, and its link with the intrinsic concentrations of nutrients and products are presented. Model parameters are fitted against experimental data, and their predictive capabilities are tested. The subsequent expansion of the model to heterotrophic cultivations is also exploited.

Chew, K. W., Yap, J. Y., Show, P. L., Suan, N. H., Juan, J. C., Ling, T. C., Chang, J. S. (2017). Microalgae biorefinery: High value products perspectives. *Bioresource Technology, 229*, 53–62. https://doi.org/10.1016/j.biortech.2017.01.006

D'Alessandro, E. B., & Antoniosi Filho, N. R. (2016). Concepts and studies on lipid and pigments of microalgae: A review. *Renewable and Sustainable Energy Reviews*, *58*, 832–841. https://doi.org/10.1016/j.rser.2015.12.162

García Prieto, C. V., Ramos, F. D., Estrada, V., Villar, M. A., & Diaz, M. S. (2017). Optimization of an integrated algae-based biorefinery for the production of biodiesel, astaxanthin and PHB. *Energy*, *139*, 1159–1172. https://doi.org/10.1016/j.energy.2017.08.036

Hata, N., Ogbonna, J. C., Hasegawa, Y., Taroda, H., & Tanaka, H. (2001). Production of astaxanthin by Haematococcus pluvialis in a sequential heterotrophic-photoautotrophic culture. *Journal of Applied Phycology*, *13*(5), 395–402. https://doi.org/10.1023/A:1011921329568

Lin, J. H., Lee, D. J., & Chang, J. S. (2015). Lutein production from biomass: Marigold flowers versus microalgae. *Bioresource Technology*, *184*, 421–428. https://doi.org/10.1016/j.biortech.2014.09.099

Shah, M. M. R., Liang, Y., Cheng, J. J., & Daroch, M. (2016). Astaxanthin-Producing Green Microalga Haematococcus pluvialis: From Single Cell to High Value Commercial Products. *Frontiers in Plant Science*, *7*(April). https://doi.org/10.3389/fpls.2016.00531

Singh, U. B., & Ahluwalia, A. S. (2013). Microalgae: A promising tool for carbon sequestration. *Mitigation and Adaptation Strategies for Global Change*, *18*(1), 73–95. https://doi.org/10.1007/s11027-012-9393-3

577731 Optimization and Scale-up of Biodiesel Production Using Lipase Immobilized on Magnetic Nanoparticles

Biodiesel production using immobilized lipase is becoming a more promising process. Lipase as a biocatalyst has several advantages over traditional base-catalyzed biodiesel production such as mild operating conditions and the option to use waste cooking oil as raw material. The performance of immobilized lipase is mostly determined by the supporting material and the method of immobilization. Lipase has been successfully immobilized on many different carriers and used in biodiesel production to test its catalytic activity and reusability. This work will focus on biodiesel produced from lipase immobilized on magnetic nanoparticles. Lipase immobilized on magnetic supports have the advantage of being recovered from a reactor by simply applying a magnetic field to the reaction slurry, eliminating the need for tedious separation procedures. This makes magnetic biocatalysts an appealing choice for industrial-scale biodiesel production from lipase. The aim of this work was to design and optimize a biodiesel production plant that uses lipase immobilized on magnetic nanoparticles, then evaluate the economic feasibility of the process.

561658 Process Design and Economic Analysis of the Production of Sucrose Esters Using a Solvent-Free Transesterification

Sucrose esters are biobased surfactants produced from sucrose (from sugar cane) and fatty acid methyl esters (derived from vegetable oils). Due to their properties such as rapid biodegradability, biocompatibility, and biocide potential for certain microorganisms [1], they are used in cosmetic, food and pharmaceutical green products.

Sucrose esters are commonly produced through base-catalyzed transesterification of sucrose and fatty acid methyl esters (FAMEs) under vacuum to remove produced methanol. Depending on the esterification degree of the sucrose molecule, sucrose esters can be produced in a wide range of polarities (i.e. Hydrophilic-Lipophilic-Balance, HLB), which is related with its final application. In the transesterification route, the reactants incompatibility is generally overcome by the use of solvents of medium polarity such as dimethylsulfoxide (DMSO) or dimethylformamide (DMF), in which both reactants are soluble [2,3]. A drawback in the use of these solvents is need of exhaustive solvent removal owing to the health concerns of DMF and DMSO, taking into account the final use of sucrose esters [1,2,4].

An alternative process to overcome the reactants incompatibility is the use of emulsifiers as contact agents in a solvent-free process [5–8]. Our last studies have been focused on the experimental study and kinetic modeling of this heterogeneous reactive system. Operational conditions regarding the optimal particle sizes (of sucrose and catalyst), temperature, reactants ratio, catalyst concentration and contact agent (emulsifier) concentration to maximize the performance of the reaction have been determined.

In the solvent-free transesterification, considerable amounts of non-reactive sucrose and FAME remain on the final mixture. The reported purification processes to obtain commercial sucrose ester propose the use of an organic non-polar solvent to extract the residual FAMEs [9–14]. Then, water or other polar solvent is used to wash the precipitated sucrose esters and to separate them from residual sucrose and catalyst. Our recent studies have been focused on the solvent selection and the determination of the operation conditions of the FAME and sucrose extraction, as well as on the product bleaching.

In this work, results of the previous works were used to propose a process design to produce sucrose esters via solvent-free transesterification. The batch operation and continuous operation were compared in two different process designs. Moreover, the process to produce three different final products: one of high HLB using potassium palmitate as contact agent (SE-KP-72), one of medium HLB using sodium stearate as contact agent (SE-NaS-51), and one of low HLB using sodium stearate as contact agent (SE-NaS-30). For all processes, equipment was selected and sized using the generated experimental information, kinetic models, empirical models and patented operation conditions. Then, correlations were used to calculate capital and operating costs, and the processes were economically assessed using two profitability indicators and the cost distribution. In general, the sucrose ester production using the solvent-free transesterification is competitive compared with the solvent transesterification. Batch and continuous processes showed similar profitability, while products using sodium stearate as contact agent had better values of the economic indicators. Obtained results can be used for the identification of the key design and operation variables affecting the economic viability of the process. These parameters should be further studied and optimized to continue with the process implementation.

References

[1] N. Otomo, Basic properties of sucrose fatty acid esters and their applications, in: D.G. Hayes, D. Kitamoto, D.K.Y. Solaiman, R.D. Ashby (Eds.), Biobased Surfactants Deterg., AOCS Press, Urbana, Illinois, 2009: pp. 275–298.

[2] P.S. Deshpande, T.D. Deshpande, R.D. Kulkarni, P.P. Mahulikar, Synthesis of Sucrose–Coconut Fatty Acids Esters: Reaction Kinetics and Rheological Analysis, Ind. Eng. Chem. Res. 52 (2013) 15024–15033. doi:10.1021/ie401524g.

[3] H.B. Hass, N.J. Summit, F.D. Snell, W.C. York, L.I. Osipow, Process for producing Sugar Esters, 2893990, 1959.

[4] Y. Shi, J. Li, Y.-H. Chu, Enzyme-catalyzed regioselective synthesis of sucrose-based esters, J. Chem. Technol. Biotechnol. 86 (2011) 1457–1468. doi:10.1002/jctb.2711.

[5] K.J. Parker, K. James, J. Hurford, Sucrose Ester Surfactants — A Solventless Process and the Products Thereof, in: J.L. Hickson (Ed.), Sucrochemistry, American Chemical Society, Washington, DC, 1977: pp. 97–114.

[6] J. Fitremann, Y. Queneau, J. Maitre, A. Bouchu, Co-melting of solid sucrose and multivalent cation soaps for solvent-free synthesis of sucrose esters, Tetrahedron Lett. 48 (2007) 4111–4114. doi:10.1016/j.tetlet.2007.04.015.

[7] H.R. Galleymore, K. James, H.F. Jones, C.L. Bhardwaj, Process for the production of a surfactant containing sucrose esters, 4298730, 1981.

[8] G.P. Rizzi, H.M. Taylor, Synthesis of higher polyol fatty acid polyesters, US 3963699, 1976.

[9] K. James, Purification of Sucrose Esters, US 4104464, 1978.

[10] S. Kea, C.E. Walker, Separation and purification of sugar esters synthesized from both aqueous and nonaqueous systems, US 4710567, 1987.

[11] S. Matsumoto, Y. Hatakawa, A. Nakajima, Process for purifying sucrose fatty acid esters, US 5008387, 1991.

[12] S. Matsumoto, Y. Hatakawa, A. Nakajima, Process for recovering unreacted sucrose from reaction mixture synthesis of sucrose fatty acid esters, US 4995911, 1991.

[13] S. Matsumoto, Y. Hatakawa, A. Nakajima, Method of producing powdery high hlb sugar fatty acid ester, US 5144022, 1992.

[14] Y. Koyama, N. Kawase, H. Yamamoto, S. Kawata, Y. Kasori, Process for producing sucrose fatty acid ester, EP 0659760A2, 1994.

567737 The Dilemma of Formic Acid Stabilization: Key Influences of Stabilizers on the Esterification Reaction

The production of methyl formate (MF) by hydrogenation of CO_2 to formic acid (FA) and subsequent esterification is a promising approach for the utilization of CO_2 as a new resource for the chemical industry.[1] Due to the positive Gibbs free energy difference of the homogeneous catalyzed hydrogenation, the addition of a stabilizer e.g. in form of an amine has a strongly positive impact on the conversion of CO_2.[2] The stabilized FA is removed from the catalytic phase into a so-called product phase. This in-situ extraction shifts the reaction equilibrium towards FA. When using methanol as a solvent for this extraction step, the product phase contains all educts needed for the esterification without further addition of chemicals. However, the influence of the choice of amine and operating conditions of the hydrogenation on the economic assessment of both reactions cannot be neglected.[3] Thus, selection of a suitable amine impacts the overall process performance and has to be investigated in detail with regard to its thermodynamic and chemical impact.

In this work, we consider the influence of different amines on a novel process for MF synthesis from CO_2 with special remark on the esterification reaction where the stabilizer of FA is present as side component. A conceptual process design was carried out in order to identify promising amines, thermodynamic properties of amines are taken into account in order to prevent relevant azeotropes and thus avoiding energy intensive separation steps. Batch experiments for the esterification reaction are performed in a 100 mL reactor and in situ spectroscopy is used to detect the change of component concentrations. Thus, the influence of the amine on the esterification reaction concerning reaction velocity and yield is investigated. However, experiments show that the influence of stabilizers on the reaction kinetics cannot be neglected. A theoretical kinetic model from literature[4] is expanded with regard to the impact of stabilization on the equilibrium constant and reaction rate. The model is validated based on the experimental results in order to define operating parameters for high time-space yields.

The evaluation of the experiments and model with relevant amines shows that all selected stabilizers have a negative effect on the yield of MF in the esterification, thus shifting the equilibrium of the esterification to the substrate side. Furthermore, the influence of temperature and feed ratio on the reaction yield are taken into account. In order to increase the product yield in the presence of amines, destabilization of the FA-amine complex has to take place. Destabilization can be achieved by exploiting the temperature dependence of the exothermal stabilization reaction. Since the esterification is an endothermal reaction, equilibrium constant of this reaction is decreasing with increasing temperature, thus, a trade-off has to be found. Another promising approach to increase the time-space yield of MF is the in situ removal of the product. Therefore, we built a reactive distillation in mini plant scale with 10 bubble trays and operated it for a given amine at moderate temperature ranges. Results show a drastically improved reaction time, improved conversion of FA as well as a product separation of MF with a purity of over 90 mol% at the top of the reactive distillation column.

Acknowledgments

We gratefully acknowledge funding for the Carbon2Chem® project by the German Federal Ministry of Education and Research (BMBF). Many thanks to Covestro Deutschland AG for providing supporting information.

References

[1] Klankermayer, J., Wesselbaum, S., Beydoun, K., Leitner, W. (2016). Angewandte Chemie Int. Ed., 55, 7296–7343. DOI: 10.1002/anie.201507458.

[2] Scott, M., Blas, Molinos, B., Westhues, C., Franciò, G., Leitner, W. (2017). ChemSusChem, 10, 1085 – 1093. DOI: 10.1002/cssc.201601814.

[3] Kaiser, T., Rathgeb, A., Gertig, C., Bardow, A., Leonhard, K., Jupke, A. (2018). Chemie Ingenieur Technik, 90, 1497-1503. DOI: 10.1002/cite.201800029.

[4] Tischmeyer, M. (2004). Ph.D. Thesis, Technical University Berlin, Germany.

572715 Thermodynamic Modeling of Kraft Black Liquor Solutions with the Electrolyte NRTL Model

The kraft pulping process is the most widely used paper pulping process across the world due to its ability to recover the necessary chemicals for wood digestion. One of the most important parts of the chemical recovery process involves the evaporation and combustion of black liquor. Black liquor solutions are comprised of a complex mixture of both organic and inorganic compounds, and due to this complexity, modeling the boiling point elevation with a rigorous thermodynamic framework is often difficult to achieve. This work focuses on using the Electrolyte Nonrandom Two-Liquid (eNRTL) model to correlate vapor-liquid equilibrium (VLE) and solid-liquid equilibrium (SLE) data to predict the boiling point elevation of inorganic and organic sodium salts and black liquor solutions. The eNRTL model connects the nonideality of the solutions with the composition by using two binary interaction parameters for each electrolyte-electrolyte and molecule-electrolyte pair in the system. Each parameter contains up to three temperature-dependent coefficients which are regressed from experimental data such as vapor pressure, osmotic and activity coefficients, liquid enthalpy, heat capacity, and solubility.

569548 An Integrated Methodology for the Design of Optimal Solvent-Based Adhesive Products with Low Toxicity

Adhesive products are widely used in consumer goods, dental composite restoration, wood processing, and in the paper and packaging, construction, and transportation industries (Ebnesajjad & Landrock, 2015). Nearly 14 million tonnes of adhesives are used worldwide to meet the market demand, which is expected to further increase in the next few years (Ceresana Market Research, 2019). As such, the manufacture of improved adhesive products that meet specified target properties has attracted increasing interest over the last decades. The design of such products, however, remains challenging as it requires (i) the combination of a large number of ingredients and different types of chemicals to obtain product formulations with desired functionalities and qualities and (ii) consideration of economic, environmental and health/safety aspects. Nevertheless, in current industrial practice, methods for identifying suitable chemical blends (number of ingredients, identity of compounds and their proportions) and corresponding processing technologies are mainly based on trial-and-error approaches. In this context, the development of systematic methodologies and tools to design adhesive formulations can be an important advantage in the field.

In this work, we present a comprehensive approach for the design of optimal adhesive products with low environmental impact (Cui et al., 2018; Jonuzaj et al., 2019). The proposed methodology integrates computer-aided design tools (Gani, 2004) and Generalised Disjunctive Programming (Raman and Grossmann, 1994) to formulate and solve the product design problem. Within this integrated approach, the number of product constituents, the identities of the components (i.e., active ingredients and solvents) and their compositions are determined simultaneously. This makes it possible to identify the best combination of active ingredient and the solvent mixture, alleviating the need to decompose the problem in sequential steps in order to specify the identity of each type of chemical. The design methodology is applied to identifying optimal acrylic adhesive formulations with low toxicity. The results show that better performance can be achieved with an integrated product design approach rather than using simplified hierarchical methods. Furthermore, significant trade-offs between competing performance objectives are highlighted through a set of Pareto optimal solutions, where different blends are evaluated based on product toxicity and concentration of the active ingredient. Finally, a ranked list of promising designs, which can serve as a guide to experiments, is generated by adding integer cuts to the integrated formulation, showing that the top-ranked products are diverse in terms of the number of mixture ingredients, the identity of the solvents and their composition.

Ceresana Market Research. Accessed 2019. Market Study: Adhesives - World (3rd edition). Available at: *https://www.ceresana.com/en/market-studies/industry/adhesives-world/ceresana-market-study-adhesives-world.html.*

Cui, J., Jonuzaj, S. & Adjiman, C.S., 2018. A Comprehensive Approach for the Design of Solvent-based Adhesive Products using Generalized Disjunctive Programming. Computer Aided Chemical Engineering, 44, pp.427-432.

Ebnesajjad, S. & Landrock, A.H., 2015. Adhesives Technology Handbook (3rd Edition), San Diego, CA, USA: Elsevier Inc.

Gani, R., 2004. Computer-aided methods and tools for chemical product design. Chemical Engineering Research and Design, 82, pp.1494-1504.

Jonuzaj, S., Cui, J. & Adjiman, C.S., 2018. Computer-aided design of optimal environmentally benign solvent-based adhesive products. Computers and Chemical Engineering. Accepted for publication.

Raman, R. & Grossmann, I.E., 1994. Modelling and computational techniques for logic based integer programming. Computers & Chemical Engineering, 18, pp.563-578.

577505 A Novel Microporous Graphite Bipolar Plate Design for Effective Water Management in PEM Fuel Cells

Proton exchange membrane (PEM) fuel cell is an attractive clean-energy technology, that could be a viable alternative to the colossal burning of fossil fuels in vehicles powered by internal combustion engines [1, 2]. The bipolar plates, which are critical components of a PEM fuel cell stack, are associated with several functions. They provide cell-to-cell electrical contact in the stack, generate uniform flow fields for hydrogen and air, act as heat transfer surfaces, and prevent leakage of gases and coolant. Bipolar plates also significantly contribute to nearly 80 % of the total weight and 45 % of the total cost of the fuel cell stack [3]. Hence, there is significant interest in finding new materials for bipolar plates. Current bipolar plates are either metal-based or graphite-based. Metal-based bipolar plates are prone to chemical attack in the corrosive environment of the fuel cell (pH of 2-4 and temperatures around 80 °C). Electrocatalyst poisoning, ion-exchange of the dissolved metal ions with the membrane, and formation of a high impedance oxide surface layer are some of the problems associated with metal plates. Graphite plates have better chemical stability than metals but are brittle. Carbon-based composite bipolar plates, on the other hand, have improved flexural strength (albeit at the cost of lower electrical conductivity) than pure graphite plates. Other characteristics such as lower density and a relatively facile manufacturing process make carbon-based polymer composites promising alternatives to metallic or purely graphite plates.

Water management is an important consideration for the effective operation of a PEM fuel cell [4]. The PEM must be sufficiently hydrated to sustain high ionic conductivity. Conversely, excessive accumulation of water, referred to as flooding of the fuel cell, is a significant problem that needs to be appropriately tackled. Water is produced at the cathode of the fuel cell due to the oxygen reduction reaction. Water is also formed by condensation from humidified reactant gas feeds. Additionally, the protons transferred from the anode to the cathode bring along their water of hydration. Because of water accumulation by these processes, the gas flow channels, feeding oxygen to the cathode, get flooded with water. Oxygen transport to the fuel cell is blocked, which results in intermittent power losses. The presence of water in the gas flow channels or the gas diffusion layer can result in the inhomogeneous and discontinuous distribution of reactants over the active catalyst area, which not only affects the performance of an individual cell but also leads to cell-to-cell performance variations within a stack. Thus, water management is a critical issue in PEM fuel cell technology.

The work reported herein focuses on the production and properties of novel graphite-polymer composite plates designed to address the water management problem [5]. Microporous plates, whose porosity was tailored for the desired up-take of water produced during fuel cell operation through capillarity, while offering sufficient resistance to permeability and leakage of gaseous fuel, were synthesized using graphite, water-based resins, and pore-forming additives. Bipolar plate properties such as water-uptake by wicking and suction (vacuum fill), the pressure required to force gas through the water seal in the plates, through-plane, and in-plane electrical conductivities, and flexural strength of the plates were measured and correlated with plate composition. Bipolar plates that met or exceeded the US Department of Energy (DOE) 2020 requirements were obtained [6]. The synthesis and properties of these new bipolar plates will be presented in this talk.

References:

1. D.F. Dominkovic, I. Bacekovic, A.S. Pedersen, G. Krajacic, The Future of Transportation in Sustainable Energy Systems: Opportunities and Barriers in a Clean Energy Transition, *Renewable and Sustainable Energy Reviews*, 82 (2018) 1823-1838. doi: 10.1016/j.rser.2017.06.117
2. A. Albarbar, M. Alrweq, Proton Exchange Membrane Fuel Cells: Review, in: A. Albarbar, M. Alrweq (Eds.) Proton Exchange Membrane Fuel Cells: Design, Modelling and Performance Assessment Techniques, Springer International Publishing, Cham, 2018, pp. 9-29. doi: 10.1007/978-3-319-70727-3_2
3. R.A. Antunes, M.C.L. Oliveira, G. Ett, V. Ett, Corrosion of Metal Bipolar Plates for PEM Fuel Cells: A Review, *International Journal of Hydrogen Energy*, 35 (2010) 3632-3647. doi:

10.1016/j.ijhydene.2010.01.059
4. O.S. Ijaodola, Z. El- Hassan, E. Ogungbemi, F.N. Khatib, T. Wilberforce, J. Thompson, A.G. Olabi, Energy Efficiency Improvements by Investigating the Water Flooding Management on Proton Exchange Membrane Fuel Cell (PEMFC), *Energy*, 179 (2019) 246-267. doi: 10.1016/j.energy.2019.04.074
5. S. Krishnan, M. Harrington, A.P. Pitchiya, Z. Putnam, D. Orlowski, Material Compositions and Methods for Porous Graphite-Polymer Composite Bipolar Plates, U.S. Patent Appl. 62/726,245 (2018)
6. A.P. Pitchiya, Microporous Graphite Bipolar Plates for Proton Exchange Membrane Fuel Cells, M.S. Thesis, Department of Chemical and Biomolecular Engineering, Clarkson University (2018).

572258 Energy-Saving Self-Entrained Heterogeneous Azeotropic Distillation for Separating Immiscible Systems

Immiscible mixtures usually form in petroleum, biofuel, pharmaceutical, chemical industries. The corresponding mixtures could be C5/methanol, *n*-butanol/water, and 2-pentanone/4-heptanone/water, etc. Any of them commonly presents heterogeneous azeotrope, so conventional distillation cannot achieve the separation completely. Fortunately, the heterogeneity of the system should be fully utilized to get the separation complete due to the formation of heterogeneous azeotropes such that it offers considerable investment reduction and energy saving. This work aims to summarize the separation of immiscible systems via self-entrained heterogeneous azeotropic distillation and use 2-pentanone/4-heptanone/water system as a case study to demonstrate the benefit potential. The design, operation and control for this common process were systematically investigated via navigating residue curve maps and balancing total annual cost and product loss.

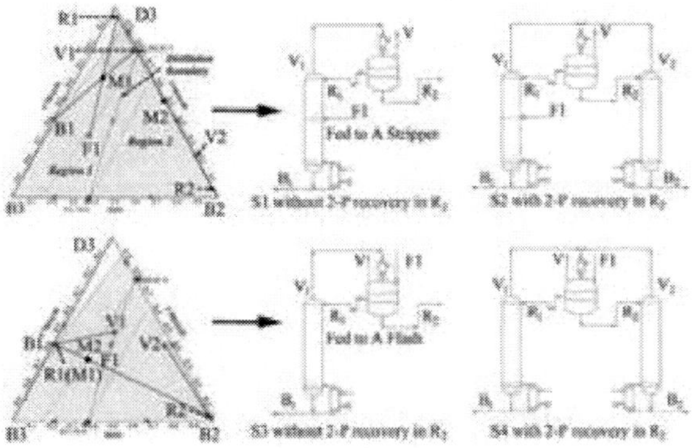

Figure 1: Ternary phase and residue curve map analyses for 2-pentanone/4-heptaone/water system and the corresponding process flowsheet: (S1) Sequence without 2-pentanone recovery in R2 and fed to a stripper, (S2) Sequence with 2-pentanone recovery in R2 and fed to a stripper, (S3) Sequence without 2-pentanone recovery in R2 and fed to a flash, (S4) Sequence with 2-pentanone recovery in R2 and fed to a flash

572846 Invited: Sensor Applications in Bioengineering: Detection of Small Molecules within Complex and Three-Dimensional Biological Milieu

Bioengineering applications in regenerative medicine and single-cell analysis present challenging microenvironments for diagnostic and real-time detection. We have developed and applied sensors to detect small molecules in neuronal, cardiovascular, blood and biofilm systems. In particular, we have aimed to establish easy-to-use sensor platforms that do not interfere with biological processes. These sensors are reagentless and can be deployed in the presence of complex macromolecular solutions that readily foul, degrade or block many traditional sensing approaches. Herein, we describe our efforts towards detection of neuroactive molecules in the brain, hypoxia gradients in three-dimensional cellular scaffolds, and subcellular oxygen concentrations during electrophysiological and chemo-mechanical single cell experiments.

575136 Tissue-like Neurochemical Sensors Operating in Brain and Gut

Normal neurotransmission relies on a precise and delicate balance of chemical signaling mediated by the release of variety neurotransmitters. Among them, the monoamines including dopamine and serotonin are uniquely involved in the regulation of cognitive processes such as emotion, arousal and memory. Therefore, it is of great interest in both basic research and clinical studies to monitor the fluctuation of monoamine in both brain and peripheral nervous systems. As a very soft and complex biomaterial, the human brain is under constant motion and deformation because of different physiological dynamics, while GI tract is a series of soft organs with inherent motility, joined in a long and twisting tube from the mouth to the anus. As a result, rigid neurochemical sensors can cause inflammation to the brain tissue, and neurochemical sensing in awake animals have historically been impossible to traditional rigid and fragile devices.

Softness and stretchability in new-generation neuroengineering tools are essential to minimize the perturbation of motions inherent in biological systems and maximize the physiological fidelity. Here we develop a new neuroengineering platform, based on soft materials, that could allow access to the brain and entire enteric nervous system for neurochemical sensing. The new neurochemical sensor is based on graphene-elastomer composite, which offers chronic, sensitive, and selective sensing of dopamine and serotonin in the brain. Ex vivo studies shows that the soft sensor could accommodate the peristalsis movement of the intestine and continuously monitor the serotonin dynamics via thermal treatment. In vivo experiments also show that the sensor could be implanted in the colon of mice to examine the drug-stimulated serotonin response. The new neuroengineering platform described here can provide unprecedented insights on the microbiota–gut–brain communication in the context of mental disorders and therapeutic interventions, and can lead to new therapies for psychiatric disorders through GI tract without invasive surgery to the brain.

570825 Cell-Based Electrochemical Sensors for Environmental Pollutant Detection

Endocrine disrupting compounds are ubiquitous; they are found in many products intended for common use, including pesticides, plastics, and even some medications. Yet, these chemicals are implicated in health problems ranging from diabetes to cancer. We have developed a cell-based electrochemical strategy to rapidly and sensitively quantify these compounds. Our strategy measures the biological activity of chemically dissimilar xenoestrogens by harnessing the specificity of a human estrogen receptor. This receptor is expressed on the surface of *E. coli* as a component of an electrochemical sandwich assay. Our sensors require no specialized skills to implement and have enabled very sensitive detection (sub-ppb levels) of environmental pollutants. We have successfully detected compounds in from complex solutions, such as baby formula, and have measured estrogenic activity from unknown compounds leeched from a BPA-free baby bottle.

577305 Biomimetic Composites As Microfabricated, Flexible, Degradable Electrochemical Sensors

Biomimetic composites as microfabricated, flexible, degradable electrochemical sensors

- M. Xu, S. Pradhan, R.K. Pal, V.K. Yadavalli

Department of Chemical and Life Science Engineering, Virginia Commonwealth University, Richmond, VA

Biomimetic composites of naturally derived and synthetic polymers provide exciting opportunities to develop mechanically flexible, physiologically compliant, and degradable bioelectronics systems. In prior work, we have shown that naturally derived silk proteins and conducting polymers (specifically poly(3,4-ethylenedioxythiophene) :poly(styrene sulfonate) (PEDOT:PSS)) can form functional bioinks for sensing and other electrochemical applications. These inks combine the mechanical strength, biocompatibility and degradability of silk proteins, with the conductivity and chemical functionalization of the PEDOT:PSS. The fabrication and characterization of photopatternable, water-based conductive inks comprising PEDOT:PSS and synthesized photoreactive silk proteins was earlier shown [1] The presence of photoreactive groups permits a fully aqueous photolithographic strategy to form conductive microelectrodes on both rigid substrates as well as flexible silk films [2].

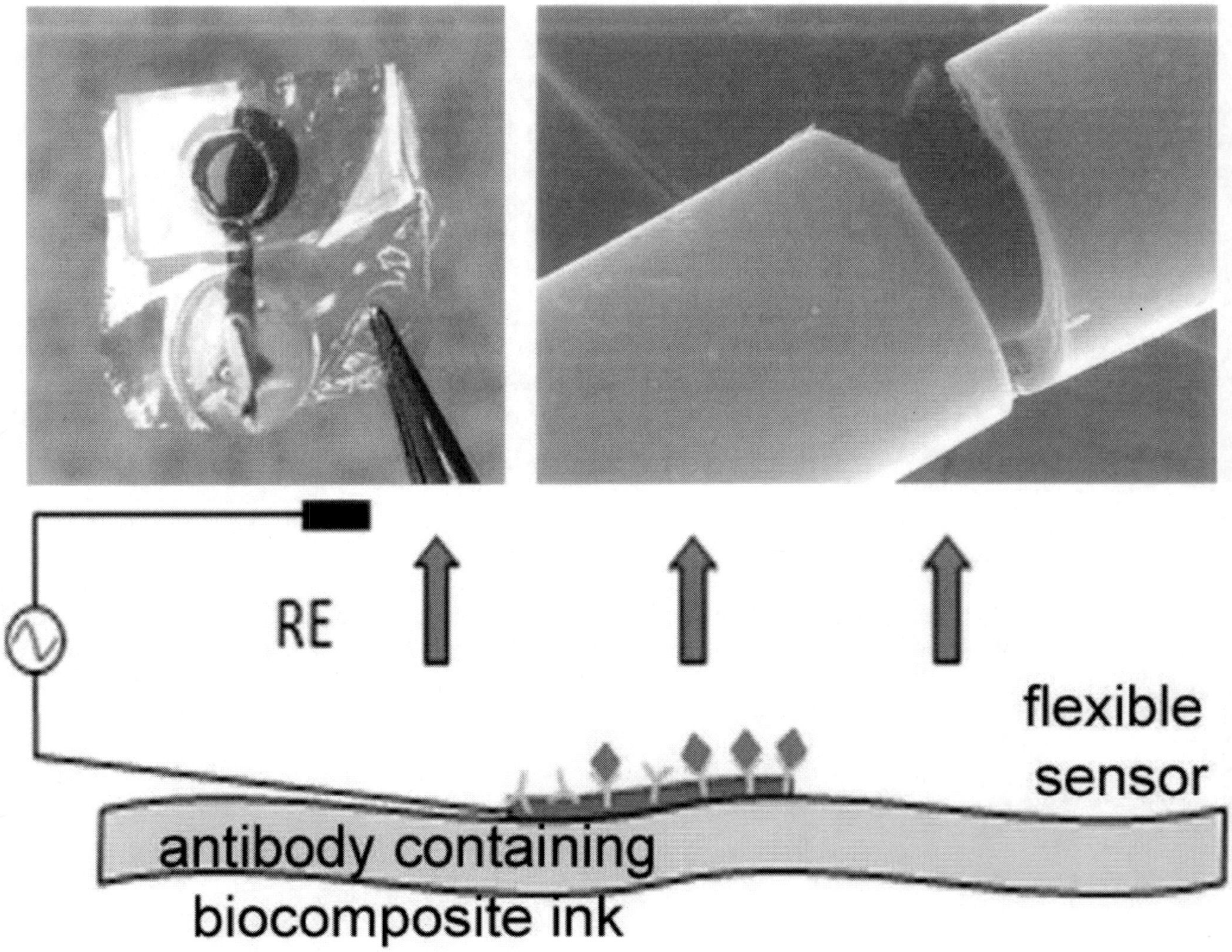

Here we will present the latest developments using this technology, showing how these devices can function for flexible bioelectronics applications in the form of microelectrodes for ultrasensitive detection, as well as components such as microsupercapacitors.[3, 4] We show how electroactive biomolecules such as neurotransmitters can be detected sensitively, while non-electroactive biomolecules such as glucose and glutamic acid, can be detected by encapsulating specific enzymes. Due to the room temperature, aqueous fabrication process, we can immobilize antibodies in these sensors resulting in flexible immunosensors for targets such as vascular endothelial growth factor (VEGF) for the first time [3]. The electroactivity of conductive ink can be improved by the addition of small amounts dopants. Using these doped composites, we further demonstrate flexible energy storage devices due to the capacitive nature of the biomaterial. The presence of silk proteins as the matrix of the composite makes it completely biodegradable and biocompatible, potentially resulting in transient or implantable devices. This platform can also be used for the specific control of cell culture in flexible sheets. The mechanical, biochemical and electrochemical characterization of the composite and its microfabrication are discussed. By virtue of a range of versatile properties, utility as bio-sensors, opto-electronic devices and flexible energy storage systems are envisioned.

References:

[1] "Photolithographic micropatterning of conducting polymers on flexible silk matrices" - RK Pal, AA Farghaly, MM Collinson, SC Kundu, VK Yadavalli, *Advanced Materials*, 28(7), 1406-1412, 2016

[2] "Conducting polymer-silk biocomposites for flexible and biodegradable electrochemical sensors" - RK Pal, AA Farghaly, MM Collinson, SC Kundu, VK Yadavalli, *Biosensors & Bioelectronics*, 81, 294-302, 2016

[3] "Flexible biosensors for the impedimetric detection of protein targets using silk-conductive polymer biocomposites" – M Xu, VK Yadavalli, *ACS Sensors*, in press, 2019

[4] "Fabrication of Flexible, Fully Organic, Degradable Energy Storage Devices Using Silk Proteins" - RK Pal, SC Kundu, VK Yadavalli, *ACS Applied Materials & Interfaces* 10 (11), 9620-9628, 2018

569635 Nanopore Based Highly Sensitive and Accurate Detection of HIV

Nanopore Based High Performance Point-of-Care Biosensor for Acute HIV Detection

Xiaojun Wei, Chang Liu

Human immunodeficiency virus (HIV) is the causative agent of acquired immuno-deficiency syndrome (AIDS). There are over 1.8 million infections HIV occur annually worldwide, and 30%¨C50% are attributed to recently infected individuals that are unaware of their disease state. Rapid diagnosis of acute HIV (within 3 weeks of infection) may reduce new cases by 10%¨C25%. The fourth-generation rapid diagnostic test for HIV is currently employed worldwide as the frontline screening test. This Enzyme-linked Immunosorbent Assay (ELISA) based test is designed to detect HIV-1 p24 antigen in addition to HIV antibodies to achieve early detection. However, several recent field tests have indicated unacceptably low diagnostic sensitivity (0%) of acute HIV-1.

Herein, a nanopore biosensing platform with single-molecule level sensing and analysis capacity has been developed to address the unmet need for rapid and accurate diagnosis of acute HIV to reduce transmission. In our approach, p24 antigens isolated from plasma samples were cleaved into several specific peptide sequences, which can be recognized and quantified by their signature signals during translocation through the nanopore. A portable QuantiPep-HIV prototype suitable for point-of-care settings was developed and tested with serum samples from HIV-infected and uninfected individuals to demonstrate the feasibility and to evaluate the diagnostic performance. Our results showed that QuantiPep-HIV can improve the diagnostic sensitivity and specificity of acute HIV.

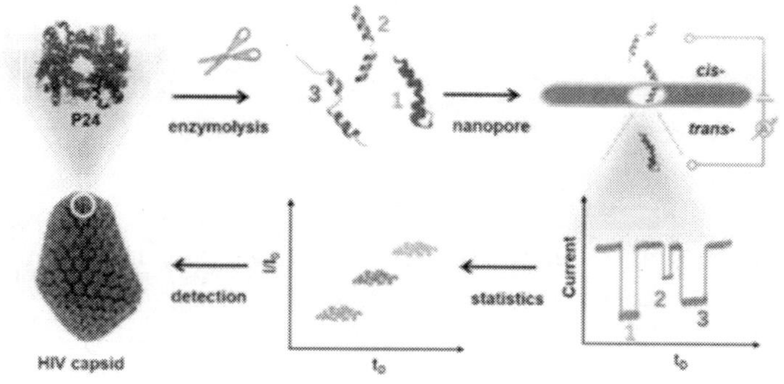

Figure. Schematic diagram for detecting peptide fragments from p24 antigens by nanopore

582378 Wireless, Soft, Skin-integrated Microfluidic Systems for Electrochemical/Colorimetric Biosensing and Capture of Sweat

Human eccrine sweat is a rich composition of nearly 60 known biochemicals including metabolites, electrolytes, vitamins, amino acids, exogeneous drugs, and small proteins. Advances in sensing technologies, conformal electronics, and materials science lay the foundation for recent demonstrations of wearable sweat sensors. Unfortunately, majority of the devices rely either on active, battery-powered electronics for electrochemical detection or passive, colorimetric chemistries for visual readout. Complex construction, large size/weight and possibility of battery leakage represent disadvantages of the former; semi-quantitative operation and limited scope of measurable biomarkers are drawbacks of the latter. Additionally, present wearable sensors can detect only a handful of biomarkers present in the sweat. Quantification of the remaining chemicals still requires sweat collection using adsorbent pads followed by off-site chemical analysis. Such a route provides average concentration of sweat analytes and restricts the acquisition of their temporal concentrations. This prohibits the understanding of the relation between dynamic variations in sweat chemistry and human physiology. In this talk, I will describe an unusual collection of enabling materials, sensing technologies, advanced microfluidics, stimulus dependent biocompatible power systems, galvanic time-tracking schemes, and wireless electronics to address these issues. Specifically, I will describe my strategies that include soft microfluidics, fuel cells, trigger-activated biocompatible batteries, colorimetrics and near field communication technologies that result in new classes of wireless, skin-integrated devices for detailed analysis of human sweat.

574301 The Electrochemical Impedance Response of a Continuous Glucose Monitor

The Gerischer impedance has been observed in systems with an electrochemical reaction coupled with a homogeneous chemical reaction. Under the assumptions of linear homogeneous reaction, equal diffusion coefficients and a semi-infinite condition, the Gerischer impedance can be solved analytically [1, 2]. However, with non-ideal electrochemical systems, where coupled nonlinear homogeneous reactions and various mass-transfer properties make exact analytical resolution impossible, a more detailed analysis is required.

In the present work, a mathematical model for the impedance response of glucose-oxidase based electrochemical biosensors has been developed [3]. The complicated coupling between thirteen homogeneous reactions and three heterogeneous reactions has been considered in the model. The homogeneous reactions included anomerization between α-D-glucose and β-D-glucose and four reversible enzymatic catalytic reactions transforming β-D-glucose and oxygen into gluconic acid and hydrogen peroxide, pH-dependent enzymatic activity and a biological buffer system. The electroactive hydrogen peroxide was considered to be reversibly oxidized or reduced at the electrode.

The mathematical model is solved numerically by using the finite-difference method and Newman's BAND algorithm [4]. The model demonstrates how the coupled non-linear homogeneous reactions effect the diffusion impedance, which has broadened the scope of the Gerischer impedance. The model can be used to explore the influence of various system parameters on limiting current, reaction profiles, and diffusion impedance. The system parameters, including interstitial glucose concentration, oxygen concentration, active enzyme concentration, diffusion coefficients, reaction rate constants and layer thickness, are related to various sensor working conditions such as body sugar level, inflammation, sensor degradation and sensor design.

References

1. H. Gerischer, "Wechselstrompolarisation Von Elektroden Mit Einem Potentialbes- timmenden Schritt Beim Gleichgewichtspotential," *Zeitschrift fur Physikalische Chemie*, **198** (1951) 286–313.
2. M. E. Orazem and B. Tribollet, *Electrochemical Impedance Spectroscopy* (John Wiley & Sons, Hoboken, NJ, 2017), 2nd edition, p. 279-293.
3. M. Gao, M. S. Hazelbaker, R. Kong, and M. E. Orazem, "Mathematical Model for the Electrochemical Impedance Response of a Continuous Glucose Monitor," *Electrochimica Acta*, **275**(2018),119-132
4. J. S. Newman and K. E. Thomas-Alyea, *Electrochemical Systems* (John Wiley & Sons, Hoboken, NJ, 2004), 3rd edition.

Acknowledgement

The support of Medtronic Diabetes (Northridge, CA) and Andrea Varsavsky, program monitor, is gratefully acknowledged.

568217 Determination and Understanding of Dynamic Lead Lag between in-Line NIR Tablet Press Feed Frame and Off-Line NIR Tablet Measurements

INTRODUCTION

Process Analytical Technology (PAT) is crucial for the understanding, monitoring and control of (continuous) manufacturing processes and essential to move towards a Quality by Design (QbD) approach for pharmaceutical drug product development [1]. Different studies have evaluated near infrared (NIR) spectroscopy for blend potency determination inside the feed frame of a rotary tablet press and compared these results with off-line tablet potency measurements [2,3,4]. The final blend first passes the NIR probe before it undergoes additional mixing inside the feeding chamber and being filled into the dies of the tablet press. As a result, the in-line NIR response is always ahead of time and more intensive compared to the signal of the material inside the tablet cores (measured by offline analysis (e.g. NIR)). This phenomenon is defined as lead-lag. Although a time-shift or lead-lag has been observed between the in-line and off-line NIR response before [2], the effect of tablet press process parameters on the observed lead-lag was not investigated and no proposal was made on how to practically use the in-line NIR potency measurements for tablet diversion. Therefore, the aim of this study was to gain an in-depth understanding of the lead-lag between in-line NIR feed frame and off-line NIR tablet measurements and to propose an approach to support the use of in-line NIR feed frame measurements for product diversion and real-time release.

MATERIALS AND METHODS Materials

For this study, test formulations containing sodium saccharine as a surrogate API were used..

Blends containing different amounts of sodium saccharine were prepared using a three-dimensional mixer (Inversina, Bioengineering AG, Wald, Switzerland).

Methods

All experiments were performed using a Modul™ P high speed rotary tablet press (GEA Pharma Systems, GEA Process Engineering, Halle, Belgium) equipped with ten 10 mm flat faced punches. All blends were compressed to a constant tablet weight of 375 mg. The NIR spectra were collected in reflection mode with a diode array spectrometer (SentroPAT FO, Sentronic, Dresden, Germany) positioned inside the feeding chamber of the feed frame, in very close proximity to the die filling station, thus monitoring blend potency immediately before the compression step [5]. Spectra were collected with an integration time of 7 ms, an averaging number of 200 spectra and a measurement cycle of 3s. A spectral range of 1100 – 2200 nm was used with a spectral resolution of 1 nm. The NIR spectra of the tablets were collected off-line by a Fourier transform near-infrared (FT-NIR) spectrometer (Antaris™ II, Thermo Fisher Scientific, USA) in transmittance mode covering the range of 7501.74 to 11998.9 cm^{-1}, with a resolution of 8 cm^{-1} and 64 scans per spectra. In-line and off-line NIR quantitative models were developed based on measurements of blends and tablets, respectively, referenced using different known sodium saccharine concentrations. Paddle and turret speeds were varied according to a full factorial design of experiments (DoE), for both model calibration and validation. The NIR models were developed using the SIMCA software (Version 15, Umetrics, Umeå, Sweden).

In order to measure the lead-lag, two series of experiments were conducted. The first series of experiments were conducted by performing blend composition step-changes (90% - 110%) of sodium saccharine during tablet compression and, the second, by adding a spike of pure sodium saccharine into the feed chute above the tablet press feed frame. To enhance the understanding of the influence of process parameters during tablet compression on the observed lead-lag; turret speed, paddle speeds, paddle speed ratio, overfill level and feed frame design were varied according to a DoE. Other equipment parameters such as paddle wheel design, position of the NIR probe and amount of punches were kept constant.

Concentration versus time profiles were plotted and lead-lag was determined.

RESULTS AND DISCUSSION

An off-line univariate linear regression NIR model using the wavelength at 8801.52 cm^{-1} and an in-line PLS model using the spectral range between 1600 and 1750 nm were obtained. The two models had an RMSEP % of 0.40 and 0.44 respectively.

From the step change experiments, it was observed that the lead-lag decreases at higher paddle speeds due to the similar responses in blend potency between NIR and compressed tablets. This phenomenon can be explained by the higher mixing intensity inside the feed frame. Similarly, the lead-lag was observed to follow the same behaviour at higher paddle speeds, when conducting spike experiments. The turret speed and overfill level were found not having a significant influence on the lead-lag within the experimental process ranges studied.

An approach was developed to use current experimental findings for supporting future use of in-line NIR for in-process control (IPC), product diversion and real-time release.

REFERENCES

[1] Wahl, P. R., Fruhmann, G., Sacher, S., Straka, G., Sowinski, S., & Khinast, J. G. (2014). PAT for tableting: Inline monitoring of API and excipients via NIR spectroscopy. European Journal of Pharmaceutics and Biopharmaceutics, 87(2), 271–278.

[2] Pauli, V., Roggo, Y., Pellegatti, L., Nguyen Trung, N. Q., Elbaz, F., Ensslin, S., ... Krumme, M. (2019). Process analytical technology for continuous manufacturing tableting processing: A case study. Journal of Pharmaceutical and Biomedical Analysis, 162, 101–111.

[3] Sierra-Vega, N. O., Román-Ospino, A., Scicolone, J., Muzzio, F. J., Romañach, R. J., & Méndez, R. (2019). Assessment of blend uniformity in a continuous tablet manufacturing process. International Journal of Pharmaceutics, 560, 322–333.

[4] Šašić, S., Blackwood, D., Liu, A., Ward, H. W., & Clarke, H. (2015). Detailed analysis of the online near-infrared spectra of pharmaceutical blend in a rotary tablet press feed frame. Journal of Pharmaceutical and Biomedical Analysis, 103, 73–79.

[5] De Leersnyder, F., Peeters, E., Djalabi, H., Vanhoorne, V., Van Snick, B., Hong, K., ... De Beer, T. (2018). Development and validation of an in-line NIR spectroscopic method for continuous blend potency determination in the feed frame of a tablet press. Journal of Pharmaceutical and Biomedical Analysis, 151, 274–283.

568340 Digitalization in Active Process Control of Pharmaceutical Continuous Manufacturing of Oral Solid Dosage

Digitalization in Active Process Control of Pharmaceutical Continuous Manufacturing of Oral Solid Dosage

Qinglin Su, Alessandra E. Anderson-Lewis, Sudarshan Ganesh, Yasasvi Bommireddy, Marcial Gonzalez,
Gintaras V. Reklaitis, Zoltan K. Nagy

Davidson School of Chemical Engineering, West Lafayette, Indiana, United States, 47907

School of Mechanical Engineering, West Lafayette, Indiana, United States, 47907

Abstract: The emerging technology of continuous manufacturing in the pharmaceutical industry requires the use of active process control strategies in order to respond to disturbances or variations in raw material properties or process parameters and thus to maintain a state of control of in-process material or product critical quality attributes (CQAs). The development of these control strategies has been identified as a need in the recent United States Food and Drug Administration¡¯s (US FDA) draft guidance for industry: Quality Considerations for Continuous Manufacturing [1]. Purdue University, together with Rutgers University, had been funded by US FDA through the project of ¡°Industry 4.0 Implementation in Continuous Pharmaceutical Manufacturing¡±, to investigate the digitalization of active process control, as shown in **Figure 1**. One of the specific aims of our work is to develop nonlinear model predictive control (NMPC) strategies based on high degree understanding of process dynamics in the product and process for the continuous manufacturing of oral solid dosage. This advanced control strategy is capable of real-time prediction of the transition of controlled variables and optimization of the control moves in the manipulated variables within a receding time horizon, so as to achieving digital process operation. The actual implementation of NMPC in pharmaceutical continuous manufacturing lines is still in its infant stage: not much progress has been as yet reported [2]. The sluggish low-level control loops were regarded as a source of medium-level common risk in our recent study [3]. Hence, a second aim of our work is to monitor the process control system itself, using performance criteria or metrics of control design analysis to identify and diagnose the root causes of degrading control performance [4], especially, in the lower-level control loops of the control hierarchy. Instead of the overall control performance evaluation based on implicit final product qualities, each individual process control loop performance can be explicitly and digitally diagnosed. The proposed digitalized active process control strategy is demonstrated on the Purdue University Pharmaceutical Continuous Manufacturing pilot plant for oral solid dosage using the direct compaction route. Specifically, a nonlinear hybrid model based on classical Kawakita model for powder compressibility is used for NMPC implementation in the continuous tablet press unit operation. The real-time model parameter estimation for Kawakita model on the basis of joint data reconciliation and parameter estimation for sensor network was recently published in our group [5]. In addition, case studies of using PiControl Apromon for digitalized control performance monitoring and OSIsoft Pi System for digitalized process performance visualization are discussed. Finally, perspectives on process model maintenance and control performance monitoring will be given.

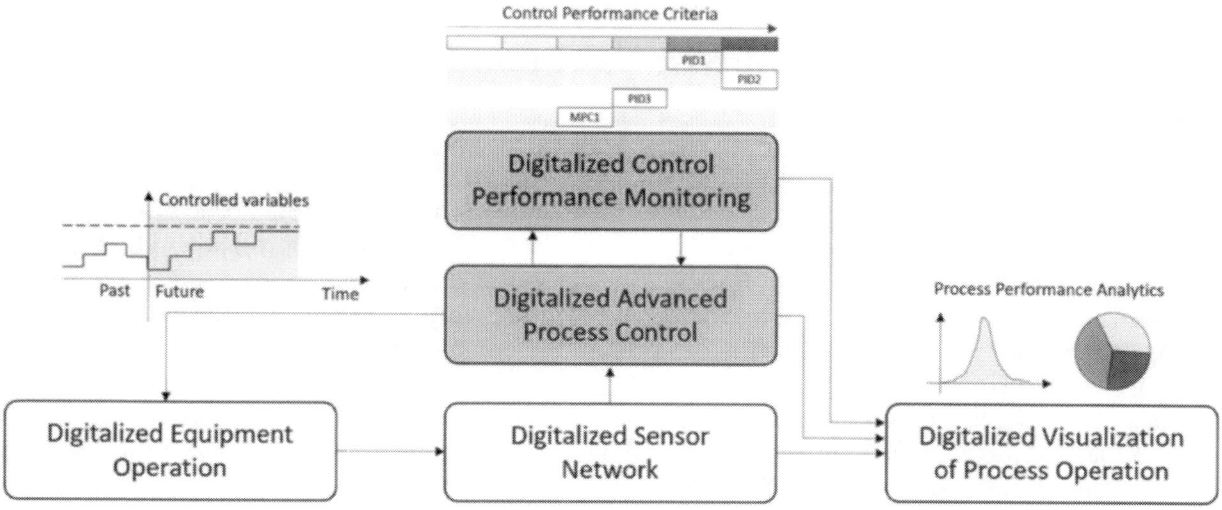

Figure 1. A schematic diagram of the digitalization in active process control.

References

1. United States Food and Drug Administration. *Quality Considerations for Continuous Manufacturing - Guidance for Industry (Draft Guidance)* 2019.

2. Mesbah A, Paulson JA, Lakerveld R, Braatz RD. Model predictive control of an integrated continuous pharmaceutical manufacturing pilot plant. *Organic Process Research & Development*. 2017;21:844-854.

3. Su Q, Ganesh S, Moreno M, et al. A perspective on Quality-by-Control (QbC) in pharmaceutical continuous manufacturing. *Computers and Chemical Engineering*. 2019;125:216-231.

4. Su Q, Moreno M, Giridhar A, Reklaitis GV, Nagy ZK. A systematic framework for process control design and risk analysis in continuous pharmaceutical solid-dosage manufacturing. *Journal of Pharmaceutical Innovation*. 2017;12:327-346.

5. Su Q, Bommireddy Y, Shah Y, et al. Data reconciliation in the Quality-by-Design (QbD) implementation of pharmaceutical continuous tablet manufacturing. *International Journal of Pharmaceutics*. 2019;563:259-272.

576091 Real-Time Monitoring and Model-Based Prediction of Product Properties in a Dropwise Additive Manufacturing Process for Pharmaceuticals

In recent years additive manufacturing (AM) technology has attracted considerable interest for its potential applications to on-demand manufacturing of dosage forms with individually adjusted doses, which is a necessity for implementation of precision medicine at the societal level, and for development of robust manufacturing platforms for early drug development or continuous manufacturing. Commensurate with the level of interest is the variety of different technologies based on two-dimensional (2D) or three-dimensional printing (3D) techniques, all of which present the common advantages of precision and flexibility in dosing inherent to AM processes. However, differences exist in the printing material requirements and in the achievable production rates for the process; both must be considered when proposing implementations in the various contexts (compounding in clinical or pharmacy setting; full-scale production). In the former, stringent requirements imply that the system would be suited to a narrow range of dosage forms, while the latter is an eminently practical consideration – if the production time for a single dose is on the order of minutes, then the system may struggle to meet demand.

Of the AM techniques, the system based on the drop-on-demand method (2D printing) for fluid deposition/dispensing has demonstrated the capacity to serve as flexible manufacturing platform for pharmaceutical dosage forms, with doses in amounts of 1-100 mg active ingredient produced from printing materials which ranged from drug solutions to drug-polymer melts and concentrated suspensions of micronized particles [1, 2, 3]. Most recently demonstrated was the development of the system for precision dispensing of micronized powders by processing them as particle suspensions, which, as a fluid-based method, avoids powder flow issues and may readily be adapted to new particulate materials using dimensional analysis [3, 4]. For dosage units which contain 100mg of active drug, production rates of up to 2000 doses per hour, with product RSD < 2%, have been achieved with suspension processing (3, 4). This system uses the drop-on-demand (DoD) printing method to deposit drops on to a prepared substrate, wherein the content of each dose is determined by the number, volume and composition of drops; consequently, quick adjustments to dosage unit properties may be made by changing the deposited fluid volume, and/or composition.

From a process perspective, the DoD method presents a unique opportunity for monitoring of dose uniformity through the online measurement of each drop added to a given substrate; this enables model-based estimation of the drug product properties (e.g. active ingredient and/or excipient amount), which when combined with statistical modeling, can be used for real-time quality control purposes. To achieve this, one must have real-time measurements of drop volume and concentration in the printing fluid just upstream of the drop ejection point. As demonstrated in previous works, drop volume measurements may be obtained using an online image acquisition system and image analysis [1, 2], with corrections necessitated by particle effects made using a Bayesian statistical model [5]. In-line concentration measurement is complicated by the small scale of the process, which precludes the use of some sensors, and by the variety of process fluids for which the sensor must have the capacity to accurately measure. For pure fluids there are few limitations imposed by physical constraints, but for more interesting printing materials such as concentrated particulate suspensions, choices for sensing technology are limited to ultrasound velocity (correlation with particle fraction through modified Urick equation) or Raman spectroscopy.

In this work we investigate these potential methods for in-line concentration monitoring for particle suspensions within the dropwise additive manufacturing system. Combining the real-time drop volume and inline concentration measurements, we are able to make model-based predictions for properties of dosage units produced during manufacturing runs for doses of variable potency; comparison of model predictions with offline analysis of the dosage units is then presented. Through the use of a statistical approach for model-based prediction, and, consequently, quantification of uncertainty in the dose content, additional modeling work enables making a decision as to whether each dose should be released to the patient, formulated based on minimization of the expected loss in the Bayesian context. In this way, an important step towards the realization of precision medicine is made, namely, model-based real-time release of doses (at least, clinical/compounding settings).

References

1. Hirshfield, L., Giridhar, A., Taylor, L. S., Harris, M. T., & Reklaitis, G. V. (2014). Dropwise additive manufacturing of pharmaceutical products for solvent-based dosage forms. Journal of pharmaceutical sciences, 103(2), 496-506.
2. Içten, E., Giridhar, A., Taylor, L. S., Nagy, Z. K., & Reklaitis, G. V. (2015). Dropwise additive manufacturing of pharmaceutical products for melt-based dosage forms. Journal of pharmaceutical sciences, 104(5), 1641-1649.
3. Radcliffe, A. J., Hilden, J. L., Nagy, Z. K., & Reklaitis, G. V. (2019). Dropwise Additive Manufacturing of Pharmaceutical Products Using Particle Suspensions. Journal of pharmaceutical sciences, 108(2), 914-928.
4. Radcliffe, A. J., & Reklaitis, G. V. (2017). Dropwise additive manufacturing using particulate suspensions: feasible operating space and throughput rates. In Computer Aided Chemical Engineering (Vol. 40, pp. 1207-1212). Elsevier.
5. Radcliffe, A. J., & Reklaitis, G. V. (2018). Bayesian estimation of product attributes from on-line measurements in a dropwise additive manufacturing system. In *Computer Aided Chemical Engineering* (Vol. 43, pp. 1243-1248). Elsevier.

578800 Process Control Strategy Development for a Challenging Drug Powder Blending Process Via Real-Time Process Analytical Technology (PAT) Monitoring and Prediction

Powder blending is one of the key unit operations, for which the performance and outcome of the unit operation could significantly impact the critical quality attributes of oral solid dosage forms such as, assay, content uniformity (CU) or uniformity of unit dose. The significance of powder blending becomes more profound for cohesive drugs and narrow therapeutic index (NTI) drugs. The following three aspects will be covered in this presentation. 1) A laboratory case study with Near-Infrared Spectroscopy (NIRS) real time process monitoring and chemometrics model prediction for a high-shear granulator powder blending process consisting of the NTI drug, Phenytoin Sodium (10% w/w) and two common pharmaceutical excipients. 2) Proper choice of the data-preprocessing algorithms to facilitate scale-up using available publicly available information. 3) Some common practices of mitigating risks to blending uniformity and content uniformity as applied to batch processing and continuous processing. This presentation will demonstrate that from both public health and process engineering perspectives, real time process monitoring, and prediction can play a vital role for understanding, controlling, and risk mitigation of a challenging drug powder blending process.

565140 Advanced Control Strategies for a Continuous Tableting Line Via Hot Melt Extrusion

From a processing point-of-view the shift to continuous manufacturing mode in pharmaceutical industry is supported by the fact that most common unit operations are continuous machinery by design, such as feeding, tableting or granulation. In-line monitoring tools to supply real-time process data and to trigger control actions are available. One remaining challenge is the establishment of real-time control concepts that enable a fully automated, yet still flexible process, capable to cope with raw material variability and process events.

In this study a feedback control strategy for a continuous tablet manufacturing line has been developed. The line consists of a hot melt extrusion and pelletizing system to produce pellets of API and polymer. The pellets are transported to a direct compaction line, blended with a pre-mix of excipients and compacted to tablets. A PAT strategy to monitor critical quality attributes of intermediates and product was implemented. Different control loops covering quality and mass flow rate were designed. PID controllers were applied for API concentration in the melt and the blend, strand temperature and strand diameter corresponding to pellet thickness. Model predictive control was used to coordinate the mass flow rates of the different unit operations. The presentation highlights an assessment of different control concepts for the line, in which the tablet press, the extruder or none act as master unit.

565007 Exploring the Robustness of Terahertz-Based at-Line Porosity Measurements for Realising a Non-Destructive Dissolution Assay of Tablets during Manufacturing

INTRODUCTION

To achieve a continuous downstream manufacturing process with in-process quality monitoring and real-time release testing capabilities, different process analytical technologies (PATs) such as laser scattering methods, near infrared (NIR) spectroscopy, and Raman spectroscopy can be deployed to, for example, measure the particle size of raw materials and powder blends as well as to detect the content uniformity of powder blends and tablets. However, the use of the above-mentioned PAT methods to directly measure and quantify the bulk properties of tablets, e.g. porosity, that govern the mass transport processes and mechanical changes in tablets during disintegration and dissolution, are limited. For example, NIR methods to quantify tablet hardness typically probes surface properties only and these measurements are not necessarily representative of the bulk tablet property, hence the need for suitable analytical technologies, such as terahertz time-domain spectroscopy (THz-TDS), that can reliably measure the porosity in transmission of tablets [1, 2]. The total porosity, f, is a more meaningful metric to measure in order to predict the dissolution performance of the tablet as it can be directly linked to the mass transport of liquid into the tablet during disintegration through physical models whilst also being predictive of the mechanical strength of the tablet ($f \quad 1/\rho$, with ρ density).

We recently developed a terahertz spectroscopy method to directly measure tablet porosity without the need for any chemometric model [1-4]. This study showcases the robustness of this method for measuring tablet porosity for a range of geometries and formulations, specifically several batches of flat and biconvex tablets with either ibuprofen or indomethacin as the active pharmaceutical ingredient (API). The study further investigates the effect of debossing on the extracted terahertz porosity. To explore the suitability of the method for possible in-line application we determined the interplay between signal averaging and error in porosity in order to determine the minimum measurement time using present commercial terahertz spectroscopy equipment. Finally, we investigate the correlation between porosity measured by terahertz spectroscopy and the mean dissolution time of immediate release tablets.

METHOD AND MATERIALS

Several batches of flat and biconvex tablets with different porosities within the range of 6 - 25% were directly compressed using a compaction simulator (HB50, Huxley Bertram Engineering Ltd, UK). The production-scale tablet press (Fette 2090 B) compression profile, with a compression speed of 60 rpm, was used for compressing all the batches. The targeted porosity for each batch was achieved by keeping the weight of all tablets at 400 mg and adjusting their thickness. The formulation used was composed of microcrystalline cellulose (39.1% w/w), lactose anhydrous Supertab21AN (46.9% w/w), croscarmellose sodium (3.0 %w/w), Magnesium stearate (1.0% w/w), and ibuprofen/indomethacin (10.0% w/w). The proportions of the various powders were weighed and properly mixed using a Turbula mixer, T2F (Willy A. Bachofen AG, Switzerland). The blending process lasted for 11 mins with a speed of 32 rpm. Terahertz time-domain measurements of all the batches were acquired using a TeraPulse 4000 (TeraView Ltd., Cambridge, UK). The porosity was extracted from the measured terahertz effective refractive index using Bruggeman's effective medium theory [1,5].

RESULTS AND CONCLUSIONS

The comparison between the terahertz porosity values measured from the debossed tablets and those without debossing showed that the presence of the debossing has a negligible effect on the results. Both flat-faced and biconvex tablets exhibited an excellent linear correlation between the nominal and the terahertz porosity. Our results show the robustness of the terahertz approach for tablet porosity measurement over a wide range of geometries. We also observed that the root mean squared error (RMSE) of the linear correlation between the nominal and measured porosity stayed $< 0.65\%$ even when reducing the signal averaging to less than one second (for comparison one minute acquisition time results in a RMSE of around 0.4%). The equipment available today therefore would be suitable for at-line porosity measurements with a potential throughput of 1 tablet per second and future improvements to the terahertz equipment appear to make on-/in-line applications possible. Finally, an excellent correlation was observed between the porosity and the mean dissolution time, which highlights the suitability of this method for non-destructive prediction of the dissolution behaviour of pharmaceutical tablets during manufacturing.

ACKNOWLEDGEMENTS: The authors would like to thank Innovate UK (Project 5804) for the funding provided.

REFERENCES

1. Bawuah, P.; Pierotic Mendia A.; Silfsten, P.; Pääkkönen P.; Ervasti, T.; Ketolainen, J.; Zeitler, J.A. and Peiponen, K.E. Detection of porosity of pharmaceutical compacts by terahertz radiation transmission and light reflection measurement techniques. Int. J. Pharm. 465, 70-76 (2014).
2. Bawuah, P.; Silfsten, P.; Ervasti, T.; Ketolainen, J.; Zeitler, J.A. and Peiponen, K.E. Non-contact weight measurement of flat-faced pharmaceutical tablets using terahertz transmission pulse delay measurements. J. Pharm. 476, 16-22 (2014).
3. Markl, D.; Strobel, A.; Schlossnikl, R.; Bøtker, J.; Bawuah, P.; Ridgway, C.; Rantanen, J.; Rades, T.; Gane, P.; Peiponen, K.-E. and Zeitler, J. A. Characterisation of pore structures of pharmaceutical tablets: A review. J. Pharm. 538, 188-214 (2018).
4. Markl, D.; Sauerwein, J.; Goodwin, D.J.; van den Ban, S. and Zeitler, J. A. Non-destructive determination of disintegration time and dissolution in immediate release tablets by terahertz transmission measurements. Res. 34, 1012-1022 (2017).
5. Markl, D.; Wang, P.: Ridgway, C.; Karttunen, A-P.; Chakraborty, M.; Bawuah, P.; Pääkkönen, P.; Gane, P.; Ketolainen, J.; Peiponen, K.-E. and Zeitler, J. A. Characterization of the Pore Structure of Functionalized Calcium Carbonate Tablets by Terahertz Time-Domain Spectroscopy and X-Ray Computed Microtomography. Pharm. Sci. 106, 1586-1595 (2017).

555745 Commercial Control Strategy Development for a High Shear Wet Granulation Process

During development of an oral drug product commercial-scale manufacturing process using high shear wet granulation, challenges were encountered due to a change in the drug substance manufacturing process and resulting API properties. These challenges required a redesign of the manufacturing process and implementation of a control strategy to ensure/demonstrate equivalency of the granules, that when processed downstream, achieved the desired drug product critical quality attributes. Appropriate PAT tools were used to facilitate development at the commercial scale. The final control strategy uses real-time data monitoring, complex recipe control and formal in-process control (IPC) requirements.

All authors are employees of AbbVie and may own AbbVie stock. AbbVie sponsored and funded the study; contributed to the design; participated in the collection, analysis, and interpretation of data, and in writing, reviewing, and approval of the final publication.

580905 Introductory Remarks for the Industry 4.0 and Digital Transformation Session

The distinguished industrial speakers will be introduced by the session chair.

579773 Leading the Industry 4.0 Journey in the Manufacturing Sector with Cross-Industry Collaboration and AI-Powered Innovation

While manufacturing represents 16% of global GDP, manufacturing industries account for 64% of global R&D spend. Yet innovation is long overdue since the manufacturing sector has experienced a decade of productivity stagnation and demand fragmentation. Organizations that have taken Industry 4.0 innovation to scale beyond the pilot phase have experienced unprecedented increases in efficiency with minimal displacement of workers. However, most companies appear to be stuck in "pilot purgatory". Widespread adoption of artificial intelligence (AI) and allied Industry 4.0 technologies at scale, through the combined efforts of companies and governments, can lead to a sizeable increase in global wealth creation, benefiting people throughout society.

However, there is potential for worker displacement if changes are not handled properly. Strengthened collaboration among stakeholders is needed to understand unexpected consequences and manage the transition.

Leaders in digital manufacturing and globalization 4.0 exhibit all the essential characteristics of the Industry 4.0 revolution. Moreover, they confirm the hypothesis that they have the potential to generate new economic value, driven by improvements in the full spectrum of production value drivers: resource productivity and efficiency, agility and responsiveness, speed to market and customization to meet customer needs. Transforming existing production systems, innovating value chains and building new business models with disruptive potential for established businesses can create value.

579424 Real-Time Analytics for IIoT

Digital transformation, connected enterprise, and IIOT are various names that refer to an increasing interest by the manufacturing industry to make better use of the data they generate. A key driver behind this heightened interest is the expectation that the business will benefit from mining actionable information out of that data.

The prevailing discourse today equates successful digital transformation with big data analytics (often delivered via a cloud-centric infrastructure). This talk will argue that, at least in industrial applications, there are a significant subset of problems for which a cloud-centered big data approach is not necessary.

This talk will present a family of embedded analytics capabilities that are designed to extract actionable information out of streaming operation data. A brief description of each capability along with the results from real-world applications of the technology will be presented.

579314 Process Optimization in the Age of Industry 4.0

Industry 4.0 represents a new trend of automation and data exchange in manufacturing technologies and fosters the concept of a "smart factory". Within this framework cyber-physical systems monitor physical processes, create a virtual copy of the physical world and make decentralized decisions. This talk provides an overview of modern optimization optimization tools and solution strategies within this distributed decision-making process. In contrast to current engineering tools which are based on simulation models and expand to optimization, we describe the development of more advanced decision-making tools that start with an optimization model and flesh out the details of the process. In particular, we highlight the development of equation oriented optimization platforms with fast solvers, advanced modeling environments and efficient interfaces between them. Moreover, we extend these features to NLP sensitivity, nonsmooth models and data-driven models. Finally, we explore the application of these tools for distributed optimization, optimization under uncertainty and the integration of design, operations and control.

580904 Value-Driven Digitalization Initiatives at Dow

Largely, the process industry has matured beyond the initial frenzy around automated reports, dashboards, and the embedding of statistics in software products, and has firmly moved toward building out infrastructure and decision-support systems to improve safety and maximize value.

Industry megatrends around M&A for narrowed domain focus, commoditization, low capital intensity, geo-political uncertainty, and emphasis on sustainability and customer experience require nimble and responsive supply chains, systems analysis for understanding trade-offs, and maximization of value from asset footprints. Further, with the emergence of AI platforms, the time has never been more ripe for practitioners to help an allegedly sluggish and slow-to-adopt industry to embed digital tools as a core part of their strategy.

A lot of emphasis has rightly been on external visibility of product portfolios to customers, as well as on internal decisions around what, when, and where to produce product---consequently involving IT, customer service/marketing, and supply chain functions. Although there continues to be significant opportunity in this space, AI technology and analytics has become increasingly relevant to R&D groups in the development of new products and innovative processes; and to maintenance and operations organizations in predicting and responding to process events.

Further, most digitalization projects fall in one of three areas of increasing implementation complexity: (1) technology that enables improved decision-making in a local environment that can be then cloned in different environments with small tweaks; (2) technology that is specific to a particular domain, business, class of products, or process; or (3) technology that spans across multiple assets and allows consolidated decision-making at scale. These three categories differ in the effort needed to standardize, scale, and get buy-in from stakeholders. These pose slightly different challenges, and the way organizations structure support organizations, empower employees, and manage programs is critical to the successful deployment of these systems and in effecting culture change.

Here, we provide specific examples where Manufacturing 4.0-related projects have enabled product, process, maintenance, and supply chain innovation at Dow, including recent success stories of deploying process intensification at scale. We also cover what Dow has done to intensify its organizations through three key capabilities---the Digital Marketplace, the Digital Operations Center, and the Digital Fulfillment Center, where domain experts and engineers sit alongside software architects, UX designers, and database experts to deliver solutions effectively. Further, we discuss how communities of practice around analytics have helped us align analytics professionals across different functions to develop a coherent strategy for the company around, for example, deployment best practices, building data pipelines, and digital twins. External partnerships with universities, consultants, and vendors do play a key role in technology transfer and adoption, and balancing the development of in-house talent with the right external support is important as well.

Through these stories, we hope to illustrate how an organization can plan a steady and firm climb out of the trough of disillusionment with digital technologies into the plateau of productivity.

580906 Q&A Panel Discussion

The distinguished speakers will hold a Q&A session with the audience, moderated by the session chairs.

559431 Group Contribution Method for Organosilicon Compounds

The prediction of physico-chemical properties of a compound from its structure plays a significant role in computer-aided molecular design (CAMD) problems. Among all quantitative structure property relationships, group contribution (GC) methods are the most commonly used. GC methods operate under the assumption that each molecular structure can be broken down into a set of functional groups where each group has a property-dependent contribution. Key molecular properties satisfy the group additivity principle and can be estimated by adding the product of the number of occurrences to the contribution of each group in a molecule, followed by a transformation function (usually nonlinear) chosen to closely match experimental data. GC methods enable the representation of a diverse chemical design space as combination of sub-functional groups. One of the most widely-used GC methods in CAMD is the one from Marrero and Gani [1,2]. This method introduces multiple levels of groups to capture the proximity effects of neighboring groups. This method has been used successfully in many molecular design methodologies, including in [3]. When it comes to silicon-containing structures, however, predictions from the GC methods are usually associated with high absolute average deviation, especially for melting point prediction. As a result, organosilicon compounds are usually excluded from computer-aided molecular design applications, despite their wide use in commercial products.

In this work, we aim to build group contribution models for silicon-based compounds to predict the following properties: boiling point, melting point, heat of vaporization, and liquid viscosity. The key to building an accurate and reliable group contribution model is to identify an optimal set of functional groups that form the basis in the representation of the properties. For each property, we use an information-criterion based model selection method to determine a unique set of optimal groups that minimize the chosen information criterion, prevent overfitting, and reduce root mean square error over the training data. Unlike most GC models that are only linear in the number of occurrences of groups, we utilize nonlinear basis functions to capture the effect of group interactions on physical properties in addition to the linear function of occurrences. We also explicitly model the contribution of a variety of structural features. A hierarchical regression method is used to determine the contributions of all present groups, interactions terms, and features. The resulting GC models are embedded in a CAMD framework [3] to exclusively generate organosilicon compounds that can be used as electronics coolants.

[1] J. Marrero and R. Gani, Group-contribution based estimation of pure component properties, Fluid Phase Equilibria, 183–184, 183–208, 2001

[2] J. Marrero and R. Gani, Group-contribution based estimation of octanol/water partition coefficient and aqueous stability, Industrial & Engineering Chemistry Research, 41, 6623–6633, 2002

[3] A. Samudra and N. V. Sahinidis, Optimization-based framework for computer-aided molecular design, AIChE Journal, 59, 3686–3701, 2013

572836 Molecular Simulations for the Design and Property Prediction of Elastomer Seals for Sour Gas Environments

Elastomers used as seals in the oil and gas industry must be able to withstand chemical corrosion by hydrogen sulfide that is present in a sour gas environment. Nitrile rubbers, which are otherwise promising in terms of resistance to oil swelling (because of the polar acrylonitrile groups) and low glass transition temperature (that makes them advantageous compared with conventional fluoroelastomers for low-temperature sealing applications), are prone to chemical modification by H_2S, resulting in severe degradation in mechanical properties. This talk will discuss the use of molecular simulations to design novel fluorinated nitrile rubbers resistant to H_2S. The solubilities and diffusion coefficients of the H_2S and CO_2 gases present in sour gas, in fluorinated HNBR elastomers, were predicted using molecular dynamics (MD). The effect of grafting density of the semifluorinated alkyl methacrylate was investigated. Solubilities calculated with MD simulation results were compared with those obtained from Grand Canonical Monte Carlo Simulations. These predictions showed the difference in the behaviors of the polar H_2S and the nonpolar CO_2. The results of molecular simulations were corroborated by measurement of mechanical properties of elastomer specimens aged in a sour gas environment at 250 °F and 1000 psig for 24 hours.

576384 Formulation of Post Cobalt CMP Cleaning Solution Based on Stability Constants of Metal Complexes

Post chemical mechanical planarization (CMP) cleaning has been challenging, with shrinking feature size, especially below 14nm. During polishing, the wafer surface is contaminated with various defects such as residual particles, foreign materials, metallic impurities, etc. One of the main challenges in post CMP cleaning is the removal of the organic contaminants (e.g. metal complexes) formed during polishing. Typical azole derivatives are used as metal passivation agents in the polishing slurries. One such commonly used passivating agent is benzotriazole (BTA), which forms metal complexes during polishing.

In this study, we focus on the formation and the removal of cobalt-benzotriazole (Co-BTA) complexes based on their stability constants. We also discuss how the various components present in the cleaning solution help in the removal of these contaminants from various film surfaces that are used in cobalt interconnects. The stability constant, log K_{eq}, of Co^{2+}-BTA complex was found to lie between 2.7 and 3.9 in the pH range of 6 ~ 12, much smaller than the Co^{3+}-ethylenediaminetetraacetic acid (EDTA) complex (log K_{eq} = 36.0). The role of potassium persulfate ($K_2S_2O_8$) and EDTA, that help in the oxidation and ligand exchange of Co^{2+}-BTA complexes, respectively, is also established. Additionally, sodium dodecylbenzenesulfonate (SDBS) was able to suppress both the potential difference (ΔE_{corr}) to about 10 mV for the Co/TiN couple, and the corrosion currents of Co film too. Thus, we were able to remove the Co-BTA complexes adsorbed on Co, Titanium Nitride (TiN), Silicon Nitride (SiN), and Silicon dioxide (SiO_2) film surfaces, by exposing them to an aqueous solution of 50 mM EDTA, 1.5 wt% $K_2S_2O_8$, and 1 mM SDBS at pH 12.

570451 Multi-Scale Approach to Evaluate Emulsification Variables (Tip velocity and Pumping Capacity) on Texture, Rheology and Stability of O/W Cosmetic Emulsions

Chemical industry has experienced major changes over past decades as a consequence of a more competitive and globalized market[1]. This has yielded an improvement in the efficiency of the operational processes in chemical facilities[2] and the development of research for designing products with higher added value[3], for instance, cosmetic emulsions. In this regard, understanding the relationships between product properties (macroscopic, microscopic and molecular), and connecting them with product formulation and process variables (e.g. energy consumption, impeller type) is a key factor for creating a new product[4][5]. According to Pradilla et al.[6], scale-up factors as the tip velocity and pumping capacity of the impeller influence the amount of energy incorporated to an emulsion, leading to changes in microscopic (drop and drop size diameter) and macroscopic properties (rheology). In turn, rheological properties can be linked to textural properties of a cosmetic emulsion[7], highlighting the need of exploring the relationships between the mixing characteristics of impellers and textural properties of emulsions. The purpose of this work is to evaluate the effect of process variables such as the pumping capacity and tip velocity, quantifying the energy incorporated during the emulsification process, on macroscopic scale and microscopic scale properties of oil-in-water (O/W) cosmetic emulsions. In this study, the macroscopic scale is represented by rheological (viscosity and storage modulus) and textural properties (firmness, consistency and cohesiveness), while the microscopic scale corresponded to the drop size distribution and mean drop size. Additionally, the effect of process variables on the emulsions long-term physical stability was assed. All emulsions were prepared using a typical cosmetic formulation, including ingredients such as emollients, humectants, thickeners, preservatives and emulsifiers. Two formulation were used throughout the study. The first one had a dispersed phase concentration of 30% (w/w), while the second one had a dispersed phase concentration of 80% (w/w). For these emulsions, the impeller type, tip velocity and the impeller pumping capacity were varied during the mixing process. Three different impeller types were used for preparing the emulsions (propeller, hydrofoil and pitch blade 45°) at two different pumping capacity levels, modifying the impeller tip velocity. In all cases, the impeller to tank diameter ratio was kept at a constant value of 0.78. Results showed that, for a given tip velocity and pumping capacity values, these scale-up factors governs the drop formation within the emulsion at the viscous regimen of flow. These results enable to discuss previous proposed ideas reported by Pradilla et al.[6] for cosmetic emulsions design and cosmetic emulsions manufacturing scale-up processes.

REFERENCES

[1] E. L. Cussler and G. D. Moggridge, *Chemical product design*. Cambridge University Press, 2011.

[2] C. Tsay, R. C. Pattison, M. R. Piana, and M. Baldea, "A survey of optimal process design capabilities and practices in the chemical and petrochemical industries," *Comput. Chem. Eng.*, vol. 112, pp. 180–189, 2018.

[3] R. Gani, "Chemical product design : challenges and opportunities," vol. 28, no. June, pp. 2441–2457, 2004.

[4] L. M. C. Sagis, "Dynamic properties of interfaces in soft matter: Experiments and theory," *Rev. Mod. Phys.*, vol. 83, no. 4, pp. 1367–1403, 2011.

[5] F. P. Bernardo, *Integrated Process and Product Design Optimization*, vol. 39. Elsevier, 2016.

[6] D. Pradilla, W. Vargas, and O. Alvarez, "The application of a multi-scale approach to the manufacture of concentrated and highly concentrated emulsions," *Chem. Eng. Res. Des.*, vol. 95, no. October, pp. 162–172, 2015.

[7] L. Gilbert, C. Picard, G. Savary, and M. Grisel, "Rheological and textural characterization of cosmetic

emulsions containing natural and synthetic polymers: relationships between both data," *Colloids Surfaces A Physicochem. Eng. Asp.*, vol. 421, pp. 150–163, 2013.

558888 Procapd: A Smart Tool for Computer-Aided Chemical Product Design

In the new global economic situation, chemical products are important options to meet the increasing requirements of modern society. Therefore, chemical-based product design and development continues to be a key topic in chemical, pharmaceutical, food and related industries as interest in high value-added products, green chemistry and product-process sustainability is receiving increasing focus [1]. However, most of the chemical-based products are still designed from empirical and knowledge-based trial and error methods even though different requirements need to be satisfied for design of different types of products. Considering the constraints of time, money and limited number of experiments that can be performed, optimal design of the product cannot be guaranteed. That is, promising product candidates may have been missed with this approach. Therefore, the development of general methods and tools to determine promising design alternatives satisfying different sets of product attributes for different product types that could serve as candidates for final selection through experiments is an option worth considering.

In this work, the latest version of a chemical product design software tool, called ProCAPD, is presented. ProCAPD is the one and only software that like the well-known process simulators, offer options for chemical product design, development and evaluation. ProCAPD helps to verify product design decisions and generates information that can be used to make design decisions [2, 3]. Like the contents of the process simulator, the product simulator needs a database of chemicals and properties, a library of models for properties as well as application process for products, numerical routines to solve mathematical problems as well as various calculation options related to product design templates. Also, like the process simulator, the product simulator comes with a user-interface to describe the problems and to obtain the simulation (analysis-evaluation) results. In order to make the chemical product simulator versatile and applicable for a wide range of problems, ProCAPD has undergone extensive improvements. It now includes a suite of databases (chemicals, solvents, aroma, color-agents and extension to ionic liquids, extended list of organic solvents, API, etc.); an extended library of property estimation toolbox (a new toolbox, ProCAPE, for estimation and analysis of molecular and mixture properties, product performance, blend compositions, environmental impact etc., has been developed); an extended design toolbox with multiple solution options for single molecule and mixture design using database search option, generate-test approach as well as mathematical optimization model (OptCAMD) [4] based option; updated design templates for single molecules, blends, formulations, emulsions, devices (for example, design templates for ionic liquids, surfactants, surrogate fuels, emotions, amino acids, etc.); and model solution strategies (database search, generate-and-test, mathematical optimization algorithms). The presentation will highlight the software architecture, the implemented computer aided methods-tools, illustrate their scope-significance through a number of new chemical product design-evaluation applications. The application examples will confirm that realistic chemicals-based product designs can be generated and integrated with focused experiments such that product development time and costs can be significantly reduced.

ACKNOWLEDGEMENTS:

The authors gratefully acknowledge the financial support from "Natural Science Foundation of China" (No. 21808025) and "the Fundamental Research Funds for the Central Universities DUT17RC(3)008".

REFERENCES:

[1] Hill, M., Boone, K., Anwar, R., Gaumer, R. B., & Camarda, K. V. (2014). The future of chemical engineering design: impact of faculty makeup and industrial needs. In Computer Aided Chemical Engineering (Vol. 34, pp. 88-97). Elsevier.

[2] Kalakul, S., Eden, M. R., Gani, R. (2017). The Chemical Product Simulator – ProCAPD. In Computer Aided Chemical Engineering (Vol. 40, pp. 979-984). Elsevier.

[3] Kalakul, S., Zhang, L., Fang, Z., Choudhury, H. A., Intikhab, S., Elbashir, N., Eden, M. R., Gani, R. (2018). Computer aided chemical product design – ProCAPD and tailor-made blended products. Computers & Chemical Engineering, 116, 37-55.

[4] Liu, Q. Zhang, L., Liu, L., Du, J., Tula, A. K., Eden, M., Gani, R. (2019). OptCAMD: An Optimization-Based Framework and Tool for Molecular and Mixture Product Design, Computers & Chemical Engineering, 124, 285-301.

578102 Digitalization to Drive Differentiation in a Hyper-Competitive Plastics Industry

We are in an environment where we are experiencing an unprecedented velocity of change in the Plastics Industry. According to a recent report, it is estimated that companies plan to invest 5% of annual revenues on digitization with expected gains of up to 25% in capacity utilization and a reduction of up to 10% in operating costs. In this talk I will draw a tie between the external drivers contributing to this change, the challenges we face and explore tools to help prepare the organization to manage through this change. Relevant examples will be shared.

576517 Risk Assessment in Unit Operations Laboratories

Incorporating chemical process safety into the Unit Operations Laboratories is the goal of six faculty and instructors from six different universities. Seven specific learning outcomes are being integrated into the curriculum, as recommended by the AICHE SAChE Committee to meet ABET requirements. One of these learning outcomes is to introduce the process of hazard evaluation and risk assessment.

Risk assessment entails identifying the frequency and the consequence of an incident. In order to identify the frequency of incidents in UO labs, two different incident reporting structures were implemented in two different universities. These reporting structures allow students, teaching assistants and faculty to report incident frequencies. Reports generated from these structures enable the programs to quantify the frequency of incident occurrences.

The other aspect of risk assessment is identifying the consequence. There are a range of consequences and these consequences must be categorized in terms of low, medium and high. While death and millions of dollars in damage may be obvious high consequences, the tolerance for risk may be much lower in an academic environment, For instance, a high consequence incident may be a student burn or damage to a $100 piece of equipment. Utilizing the collective knowledge of the group, a consequence scale was established.

This presentation will present details on the reporting structures and a risk matrix that may be used for university unit operations laboratories.

552669 Using Simulators for a Competitive Process Safety Exercise

Columbia University Chemical Engineering and Simulation Solutions Inc. have developed a competitive process safety chemical engineering exercise using simulators in order to educate students on chemical engineering process safety. As part of a four-day Chemical Engineering Operations Center educational exercise combining computer simulation and classroom instruction this team educates students on chemical engineering process safety and provides a unique chemical engineering process safety experience. In 2019, this academic and industry team developed a new method of using the simulators inspired by Esports. Multiple years of integrating simulation of chemical engineering processes, the CHEN E4501 Chemical Engineering Process Safety Course uses case studies, computer aided design, and simulation in the course content covering instructional objectives. The results of the 2019 lesson sequence and the Competitive Process Safety Exercise sponsored by Simulation Solutions Inc. and the methods used to evaluate student performance as related to safety outcomes are reported on here. This presentation will focus on the competitive exercise results.

578885 Classroom Use of Process Simulators to Quantify Flashing Two Phase Release Conditions

Process simulators are a readily available tool for making complex thermodynamic calculations which are necessary for quantifying the conditions for flashing two-phase flow under sudden depressurization. Many common toxic industrial chemicals (TICs) such as chlorine and ammonia are stored as liquids at ambient temperature, and these materials will flash when depressurized to atmospheric pressure forming an aerosol. This paper considers how to use a process simulator to estimate the mass flow rate from a simple hole for flashing two-phase flow. Previously published data are considered to demonstrate the effectiveness of this approach. The result of this work is the development of a general approach for quantifying the mass release rate of flashing two phase releases for TICs and other flashing liquids including mixtures with supercritical components which cannot use current textbook methods.

557652 Chemical Engineering Student Moral Reasoning within Hypothetical Process Safety Contexts

In light of recent industrial process safety incidents, ABET requires accredited undergraduate chemical engineering programs to train students to make effective and accurate process safety decisions. While many programs train their students to effectively identify and evaluate risks relevant to process safety, the continued occurrence of incidents indicates that elements of the environment in which engineers are tasked with making decisions may not be effectively captured through current instructional methods. As such, we have examined how senior chemical engineering students apply moral reasoning when faced with process safety decisions in two contexts: a text-based assessment instrument and a virtual process safety environment.

In the text-based assessment, students took the Engineering Process Safety Research Instrument (EPSRI) which consists of 5-7 process safety dilemmas and decisions. Students are then asked to rate the importance of 10-12 considerations, unique to each dilemma, to their decision-making process. In the second method, a virtual environment was developed to simulate a chemical plant, including safety hazards, interactions with co-workers, and other authentic constraints. After certain decision points, students were asked to rate how important certain considerations were to those decisions. Based on the responses to the considerations on the EPSRI and the virtual environment, students' moral reasoning could be determined by the ratings they assigned to considerations that were pre-conventional (focused on self), conventional (focused on immediate contacts such as family, employees, and company), and post-conventional (focused on the greater community) as outlined in Kohlberg's moral development theory. Comparisons of students' moral reasoning between the text-based EPSRI and more authentic environment of the virtual simulation will be discussed in the context of behavioral ethics, examining how the different scenarios impact ethical behavior and decision-making.

553646 The 4th Edition of Chemical Process Safety, Fundamentals with Applications

It has been almost 30 years since the first edition of *Chemical Process Safety, Fundamentals with Applications* was published in 1990. The second edition was published in 2002 and the third edition in 2012. The fourth edition will be published in May of 2019 and is a major revision - several chapters were completely rewritten and almost all chapters heavily revised.

This paper will discuss the original intent and history of the textbook. In addition, it will present details on the 4th edition, including new content that was added and content that was removed. We believe the 4th edition realigns the textbook with current process safety concepts and practices and provides an even more meaningful learning experience for engineering undergraduates.

579268 A Review and Outlook on Machine Learning and Data Science in Chemical Engineering

With today's development of the internet, internet of things, wireless sensors and communications, smart manufacturing, and high-throughput experiments, the amount of data collected has grown exponentially, analogous to the Moore's law in the increase of solid-state transistor density. The data explosion has made all sectors including engineering, medicine, business, commerce, finance, and materials science to endorse the potential and power of big data. The relevance of big data to chemical engineering and science is evident from the recent technical contributions by chemical engineers and scientists.

Great successes in machine learning today have to do with natural language processing (NLP), multi-lingual translation, image recognition, etc. In addition, the demonstration of AlphaGo by Google's DeepMind in early 2016 was one of the most shocking events to the world. However, the game of Go has little in common with the degree of uncertainty and the complexity of real-world engineering problems. In this talk we discuss the characteristics of process systems engineering problems and provide a brief review of machine learning and data analytics in chemical engineering in the last three decades. We give a few attributes for machine learning techniques to be widely applicable in the process systems. Owing to these characteristics, process monitoring and inferential sensors are two of the most adopted data-driven applications in chemical engineering.

The talk will then provide an assessment of the resurgent interest in deep neural networks and deep reinforcement learning and contrast them against the technology two decades ago. We also give a few examples of high-profile applications that failed the test of time. We assert that most of the sustainable applications are domain-specific, in that they incorporate specific structures and knowledge in the models. We will briefly touch on the tremendous advancement in statistical machine learning, which are the core of data science and could propel the advances in other scientific and engineering disciplines, such as smart manufacturing, bioinformatics and materials science.

559766 Neural Network-Based Modeling and Operation for ALD of SiO_2 Thin Films Using Data from a Multiscale CFD Simulation

Atomic layer deposition (ALD) plays an important role in the manufacturing of highly conformal thin-film materials, which are used extensively in the semiconductor industry. In our previous study, a multiscale computational fluid dynamic (CFD) model has been developed for the SiO2 ALD process in an optimized showerhead ALD reactor geometry. The optimized reactor design involves a volume adjusting horn in the upstream region and a radially adaptive sizing of the showerhead holes [1], [2] . The results from the developed macroscopic CFD model show that those two design features enable quick achievement of steady-state in the gas-phase domain of the reactor, and provide a relatively uniform surface pressure and temperature profile for the microscopic kinetic Monte-Carlo (kMC) modeling of the surface deposition, which authentically simulates the film deposition process [3].

However, the multiscale CFD model is computationally demanding and is not suitable for applying real-time control schemes like model-based controllers (MPC). Motivated by this, in this work, we investigate the transient operation of the full reactor in details, using the model developed previously. The impact of operation parameters, including inlet process gas flowrate, temperature and pressure, on the deposition process is discussed and a corresponding database with temporal information of the process is constructed. Also, the appropriate operating boundary accounting for both microscopic and macroscopic operations is determined to ensure process feasibility [3]. Subsequently, a recurrent neural network (RNN) is constructed and trained using the model-generated database, which enables the prediction on the transient process dynamics based on the inputs of operating condition [4]. Moreover, utilizing the RNN model, a control scheme is designed based on the developed RNN to reduce the impact of potential disturbances on the deposition process and enhance the film uniformity. Finally, an operation example is shown to demonstrate the practicability of the proposed method.

[1] Lee, C.S., Oh, M.S., Park, H.S. US Patent No. US7138336B2. 2003. Retrieved from

https://patentimages.storage.googleapis.com/01/e8/3e/5c751ffe023e4d/US7138336.pdf

[2] Crose, M., Zhang, W., Tran, A., Christofides, P.D. Multiscale three-dimensional CFD modeling for PECVD of amorphous silicon thin films. Computers & Chemical Engineering. 2018, 113, 184 – 195.

[3] Ding, Y., Zhang, Y., Kim, K., Tran, A., Wu, Z., Christofides, P.D. Microscopic Modeling and Optimal Operation of Thermal Atomic Layer Deposition. Chemical Engineering Research and Design. 145, 159-172, 2019.

[4] Connor, J.T., Martin, R.D., Atlas, L.E. Recurrent neural networks and robust time series prediction. IEEE Transactions on Neural Networks. 1994, 5, 240-254.

579520 Keynote Talk: Intelligent Cloud-Based Algorithms for Pharmaceutical Drug Development

The ability to predict or explore polymorphism in organic crystal compounds is of great importance for many industries since many physical and chemical properties of molecular crystals strongly depend on the molecular packing in the solid state. Crystal structure prediction (CSP) aims to reduce the risk and improve the efficiency of pharmaceutical drug manufacturing by correctly predicting the low energy crystal forms of organic molecules from minimal information, such as the chemical diagram. At XtalPi, we have developed a cloud-based computational platform that combines advanced physics-based algorithms with A.I/machine learning algorithms. Our CSP platform can rapidly generate and rank millions of crystal structures for pharmaceutically relevant systems and we highlight some cases in which our CSP platform has revealed low energy polymorphs in challenging systems. We will also discuss some areas where we are expanding our computational platform to make accurate predictions on the physiochemical and pharmaceutical properties of small-molecule candidates for drug design, solid-form selection, and for other critical aspects of drug development.

568479 Industrial Success Stories with Machine Learning and Deep Learning in Matlab

Machine learning has rapidly gained in popularity, spurred by well-publicized advances like deep learning and widespread commercial interest in big data analytics. Despite the enthusiasm, companies and researchers are struggling to deploy machine learning models in chemical plants so as to reconcile the hype with actual application and success.

In this presentation, MathWorks will showcase the successes of machine learning and deep learning in industrial settings with 6 user stories/case studies - all drawn from the process industries. With these examples drawn from customers' usage of MATLAB, MathWorks seeks to dispel the notion that machine learning is solely the subject of academic study but can indeed be applied to process plants with measurable business benefits.

In addition, machine learning workflow elements will be discussed in this talk including pre-processing, model building using apps and deployment to industrial systems using MATLAB.

569189 Keynote Talk: Modern Data Analytics to Accelerate the Design of Advanced Materials

We will discuss three critical aspects of applying a machine learning approach to accelerate the design of advanced materials for energy applications: 1) consistently measured, relevant experimental data with known pedigree, 2) in-depth scientific features that can be correlated with target properties, and 3) training of surrogate machine learning models. We will present examples of predicting mechanical properties and oxidation response of high-temperature alloys, thermochemistry of complex oxides, and melt-pool microstructure of single-track by the additive manufacturing process.

The research was sponsored by the LDRD Program of Oak Ridge National Laboratory and the Department of Energy, Vehicle Technologies Office, Propulsion Materials Program.

570365 GPU-Accelerated Computation of Stirred Tanks

A fast and economical solver, accelerated by the Graphics Process Units (GPU) of a single graphics card in a desktop computer, is developed for the simulation of stirred tanks, integrating Lattice Boltzmann Method (LBM), coherent Large Eddy Simulation (LES) and Immersed Boundary Method (IBM). The grid resolution can reach 13.8 million nodes in maximum, resolving considerable details of flow field in stirred tanks with only such a simple desktop computer. In the meantime, the computational speed is 1500-fold faster than that of the traditional transient simulation of Computational Fluid Dynamics (CFD) based on Navier-Stokes equations implemented on 16 CPU cores. It only takes less than two minutes to update the simulation of one impeller revolution, enabling the transient simulation of longer physical time of stirred tanks. We find that at least 4,000 impeller revolutions in simulation is needed to achieve an obviously time-independent macro-instability frequency. With the coherent LES model and reasonable grid resolution, the fast solver is able to resolve more than 95% of total Turbulent Kinetic Energy (TKE) for highly turbulent region, reproducing the monotonic decrease of TKE along the radial direction, much better than the Smagorinsky model. The average velocity and TKE predicted by the fast solver are in better agreement with the experimental data in literature. The solver is therefore more suitable for the fast simulation of stirred tanks using only a desktop computer without the need of much finer grid resolution and supercomputers. We believe this high-resolution approach serves as a good gateway to further integrate with Artificial Intelligence for chemical reactors.

564376 Keynote Talk: Faster Prediction with Artificial Intelligence: Using Process Simulations and Production Data to Develop Fast-Running Inference Models for Manufacturing

Navigating a large-dimensional design space with simulations is a significant improvement over doing so with physical experiments but can still be very time consuming. Deep learning methods are often used to develop fast-running surrogate models of more computationally expensive simulations in order to get near-real-time predictions and to optimize complex systems such as manufacturing processes, energy systems, and fusion reactors. In this talk, I will discuss an application of machine learning to develop a fast-running surrogate models that captures the dynamics of industrial multiphase fluid flows. I will also discuss an improved population search method that can help the analyst explore a high-dimensional parameter space to optimize production while reducing the model uncertainty. Results from demonstration problems and from real-world Inertial Confinement Fusion simulations will be presented.

580106 A Perspective on Teaching Process Safety through the Curriculum

It is important to properly place process safety topics in the chemical engineering curriculum to emphasize the importance of the topic and to introduce students to the key concepts. One of the best approaches, in principle, is to cover chemical process safety topics in most, if not all, core courses in the curriculum. In this paper, we will discuss the pros and cons, as well as give example of how this might be accomplished.

574468 An Initiative on Safety across the Chemical Engineering Curriculum Scott Fogler

Sometimes chemical process safety is taught in a separate safety course within the chemical engineering curriculum, and sometimes it is taught only in the senior year as a part of the process design course. The purpose of this initiative is to provide faculty and students with real case studies and resources so that process safety can be more effectively and easily learned throughout the curriculum and become an integral part of chemical engineering culture.

To achieve this culture, safety modules have been developed to be used as homework problems in every core chemical engineering lecture course (http://umich.edu/~safeche/). Each module consists of viewing a Chemical Safety Board Video, filling out an analysis of the accident, doing a course specific calculation, assigning the NFPA symbols and filling out a Bow Tie Diagram for the accident. For example, in the chemical reaction engineering course, the T2 laboratories accident and the ExxonMobil Refinery Fire are examples presented and analyzed. In addition to the core lecture courses, six video snippets (2-5 minutes each) on Laboratory Safety are included on the website.

In addition to the modules, tutorials are given on such things as the NFPA diamond, the Bow Tie Diagram, the Fire Triangle, the Process Safety Triangle and A Safety Analysis of the Incident. The solutions to the homework problems are also given on the website, but instructors will have to email the author to receive the password to enter the solutions.

The website is free and accessible to all.Sometimes chemical process safety is taught in a separate safety course within the

chemical engineering curriculum, and sometimes it is taught only in the senior year as a part of the process design course. The purpose of this initiative is to provide faculty and students with real case studies and resources so that process safety can be more effectively and easily learned throughout the curriculum and become an integral part of chemical engineering culture.

To achieve this culture, safety modules have been developed to be used as homework problems in every core chemical engineering lecture course (http://umich.edu/~safeche/). Each module consists of viewing a Chemical Safety Board Video, filling out an analysis of the accident, doing a course specific calculation, assigning the NFPA symbols and filling out a Bow Tie Diagram for the accident. For example, in the chemical reaction engineering course, the T2 laboratories accident and the ExxonMobil Refinery Fire are examples presented and analyzed. In addition to the core lecture courses, six video snippets (2-5 minutes each) on Laboratory Safety are included on the website.

In addition to the modules, tutorials are given on such things as the NFPA diamond, the Bow Tie Diagram, the Fire Triangle, the Process Safety Triangle and A Safety Analysis of the Incident. The solutions to the homework problems are also given on the website, but instructors will have to email the author to receive the password to enter the solutions.

The website is free and accessible to all.

561465 Learning Hazards of Exothermic Reactions Via Real and Virtual Experiments

Although there is a recognized need for chemical reaction safety education, there is a dearth of published studies on experiments designed to demonstrate hazards of exothermic reactions. We developed an experiment to study scale-up of an exothermic reaction and provide students with dramatic first-hand knowledge of the hazards of thermal runaway. This experiment has been implemented successfully in our unit operations laboratory and in that of at least one other university. In this presentation we discuss a way for a broader audience of students to safely experience thermal runaway via a virtual experiment.

We developed a computer simulation for the iodide catalyzed decomposition of hydrogen peroxide using COMSOL Multiphysics® software. The simulation was converted to a user-friendly "app" that allows a scale-up study of this reaction by manipulating variables in a control panel and observing temperature and pressure profiles as the reaction proceeds under different conditions. Using COMSOL Server™ software we have provided access to the app to anyone with an internet connection and a web browser. The app provides a virtual experiment that requires no knowledge or installation of the COMSOL Multiphysics® software. It is anticipated that interacting with the app will provide students with a lasting understanding of the consequences of the fact that reaction rate, and concomitant temperature and pressure increases, scales with reactor volume while necessary heat removal typically scales with reactor surface area.

562663 Fundamental Approaches to Reactive Chemicals Hazard Recognition and Analysis

Over the past half century, Dow's Reactive Chemicals Program has supported a strong safety culture and a scientifically rigorous, risk-based review process for chemical hazards in the company's labs and plants. The foundation of Dow's chemical and process hazard analyses is the data generated by the Reactive Chemicals Group, who leverage measurement science, expertise in chemistry and engineering, and an understanding of Dow's processes to support safe operation at all scales. In spite of the critical role the Reactive Chemicals discipline plays in safe chemical research and manufacturing, these approaches and techniques are absent from the majority of chemistry and chemical engineering undergraduate curricula. In this talk, we discuss the fundamental principles behind assessing chemical reactivity hazards. We focus on material that could be incorporated into the undergraduate curriculum, such as assessing theoretical and experimental heats of reaction, calorimetric techniques for studying both desired and undesired reactions, and recognizing flammability and dust explosion hazards. Incorporating reactive chemicals testing or modeling into an undergraduate curriculum would satisfy expectations and requirements for safety instruction and better prepare students for careers in both industrial and academic roles.

578267 New Liquid Flammability Rating with the Assist of Machine Learning

Currently, the classification and rating of liquid's flammability is based on its flashpoint. However, flashpoint becomes insignificant when the liquid is present in the form of aerosol. This study uses DIPPR 801 database to find a new liquid flammability rating with the consideration of the possibility of atomization. Over 800 organic chemicals have been studied and principle component analysis is applied. Also, several unsupervised learning algorithms have been utilized to conduct the clustering on the database. The new liquid's flammability rating standard will extend the existing liquid flammability standards, such as NFPA 704 and GHS of Classification and Labelling of Chemicals.

575192 Formulating Biologics in Polymeric Nanoparticles: New Methods to Overcome Manufacturing Barriers

Nanoscale polymeric vehicles can protect biologics such as peptides and proteins from clearance and produce a sustained release profile.1 The double emulsion process to form polymeric microparticles and nanoparticles is a commonly employed formulation technique but has well-documented trade-offs.2 This results in processes that are not economically viable because significant therapeutic quantities are lost as unencapsulated material. To achieve the same therapeutic dose with a low-loading formulation requires large excipient amounts and possibly high-volume injections.

We have developed a process that addresses these shortcomings by replacing the primary emulsion with an "inverse nanoparticle" containing the biologic. We employ inverse Flash NanoPrecipitation (iFNP) to produce these polymeric nanoparticles encapsulating a biologic at high loading (>50%).3,4 These inverse nanoparticles reduce process losses during production of the second emulsion due to the presence of a hydrophobic polymer shell around each biologic core. Alternatively, we have developed a process to introduce a dense PEG coating on the inverse nanoparticles to produce long-circulating water-dispersible particles structurally analogous to polymersomes or vesicles.

We detail the methods employed in forming these particulate delivery systems for sustained release applications. The loadings and encapsulation efficiencies (EE) achieved are in many cases an order of magnitude better than existing techniques. We will report approaches for tuning releases profiles and describe potential applications of the technology. The table below highlights process metrics for a range of biologics encapsulated in coated nanoparticles formed by iFNP.

Biologic	MW (g/mol)	EE (%)	Loading (%)
Lysozyme	14,400	97.7 ± 0.4	31%
Polymyxin B	1,302	80.4 ± 1.6	29%
HRP	44,000	63.7 ± 6.5	20%
MPEP-1	~3,410	29.8 ± 2.4	6%
Vancomycin	1,449	11.8 ± 3.9	2%

(1) Yu, M.; Wu, J.; Shi, J.; Farokhzad, O. C. Nanotechnology for Protein Delivery: Overview and Perspectives. J. Control. Release 2016, 240, 24–37.

(2) Pagels, R. F.; Prud'homme, R. K. Polymeric Nanoparticles and Microparticles for the Delivery of Peptides, Biologics, and Soluble Therapeutics. J. Control. Release 2015, 219, 519–535.

(3) Markwalter, C. E.; Prud'homme, R. K. Inverse Flash NanoPrecipitation for Biologics Encapsulation: Understanding Process Losses via an Extraction Protocol; 2017; pp 275–296.

(4) Pagels, R. F.; Prud'homme, R. K. Inverse Flash NanoPrecipitation for Biologics Encapsulation: Nanoparticle Formation and Ionic Stabilization in Organic Solvents; 2017; pp 249–274.

572141 Co-Immobilization of Enzyme Cascades Enables One-Pot Industrial Immobilized Biocatalysis

Enzymes are increasingly used to access novel chemical spaces and enable regio- and stereo-selective synthetic routes for a variety of drug substances. In many instances, biocatalytic transformations require concurrent operation of multiple enzymes to perform co-factor recycling or to enable cascaded transformations. While such transformations enable novel chemistry, they also increase the protein burden, which must be managed to both enable product isolation and mitigate potential off-target interactions from host-cell proteins which accompany unpurified recombinantly expressed proteins.

Immobilizing an enzyme to a porous resin solves many of these issues by creating a solid-supported catalyst with extremely high enzyme loads that can be prepared in advance, easily charged into a reactor, and rapidly isolated from product by filtration. We have developed a scalable enzyme-immobilization platform technology to selectively co-immobilize enzyme cascades onto a single resin directly from crude cell lysate and reject host-cell proteins. We have successfully utilized this technology to enable numerous multi-enzyme chemistries and applied it across multiple scales. In this talk we will discuss development and scale-up of multi-enzyme immobilized processes and the unique opportunities they afford for industrial immobilized biocatalysis.

573823 Application of Membrane Technology for Metal Removal from a Reaction Mixture

Transition metals are known for their good catalytic properties. Commensurate with the reputation, their use in various chemical transformations has gone up in recent years. However, their use in homogeneous catalysis in reactions often results in high metal content in the product. For many industries, especially for active pharmaceutical ingredients (APIs), such a high level of metal content cannot be tolerated, and the reaction mixture or product needs to be treated for metal removal.

Depending on the amount of metal contaminant, traditional, adsorption-based technologies may suffer from high adsorbent charge, long cycle times, and high costs. In this study we evaluate the application of membrane technology to remove metals from different reactions mixtures. In this presentation, we summarize the engineering methodology to develop a membrane process and touch upon scale-up considerations and challenges for successful technology implementation in a cGMP chemical pilot plant setting.

Mike W. Rasmussen, Manish S. Kelkar, John Hartung, Harsit B. Patel, Thaddeus S. Franczyk, Onkar Manjrekar, Geran Zhang, Elie Chaaya, and Nandkishor K. Nere are employees of AbbVie. The design, study conduct, and financial support for this research was provided by AbbVie. AbbVie participated in the interpretation of data, review, and approval of the publication.

574496 Applications of Soft Sensors in Pharmaceutical Manufacturing Processes

Drying is an essential processing step for residual solvent reduction in pharmaceutical processes. For the manufacture of Active Pharmaceutical Ingredient (API), drying is mainly used to produce crystals with the right form and/or particle size distribution that are suitable for down-stream processing. As an energy-intensive and time-consuming process, drying is usually a bottleneck of pharmaceutical manufacturing processes. Therefore, it is critical to well understand the drying process and improve the process efficiency. For the experimental characterization of drying progression, loss on drying (LOD) is a commonly used off-line measuring approach. LOD requires to take samples periodically and to weigh the loss in mass when the sample is continuously being dried until the weight becomes constant. However, LOD is a time-consuming and intrusive method that can disrupt the drying process and increase the cycle time. It is thus desired to develop on-line measurements to monitor the drying process.

Process Analytical Technology (PAT) has been developed as an in-line or on-line monitoring approach to provide real-time data about the process status and critical process attributes (CQA), which can facilitate the process control and optimization. Spectroscopy methods, such as near infrared (NIR) and Raman, have been widely adopted to enable the implementation of PAT in monitoring the removal of solvents in the drying process [1, 2]. However, such methods require the development of chemometric models, which typically requires large amounts of experimental data for calibration and validation. Additionally, mass spectrometry has been developed to predict the drying end point without the need of chemometric models [3]. The common feature of the mentioned PAT methods is that they usually require the installation of spectroscopic devices which can be prone to fouling and introduce additional capital costs [4].

Recently, soft sensors have emerged as a powerful tool to provide real-time process monitoring capabilities in many industry fields. Soft sensors are mathematical models that use easily accessible on-line data to estimate other variables of interest that can only be determined at low sampling rates or through off-line analysis [5]. For pharmaceutical processes, soft sensors have great potential to either replace the existing hardware sensor or work in parallel to provide redundancy and verify whether the hardware sensor is working properly [6]. However, applications of soft sensors in the pharmaceutical industry are still limited, partly due to the challenge of validation by regulatory agencies [7].

In this work, we proposed a soft sensor based on a hybrid modeling approach to estimate the product LOD in real-time during the drying process. The model combines first-principle relationships with empirical correlation models derived from historical data, which is more robust than data-driven models in handling variabilities in raw material and process conditions. The developed soft sensor has been successfully calibrated, validated with experimental data, and implemented in an API full scale commercial process that follows good manufacturing practices (GMP) procedures.

1. Hamilton, P., et al., *Studies of particle drying using non-invasive Raman spectrometry and particle size analysis.* 2011. **136**(10): p. 2168-2174.
2. Morris, K.R., et al., *Accelerated fluid bed drying using NIR monitoring and phenomenological modeling.* 2000. **26**(9): p. 985-988.
3. Dodda, A.G., et al., *Using online mass spectrometry to predict the end point during drying of pharmaceutical products.* 2014. **19**(1): p. 122-131.
4. Pla, D.L., et al., *Moisture soft sensor for batch fluid bed dryers: A practical approach.* 2018. **326**: p. 69-77.
5. Luttmann, R., et al., *Soft sensors in bioprocessing: a status report and recommendations.* 2012. **7**(8): p. 1040-1048.
6. Kourti, T.J.A. and b. chemistry, *The process analytical technology initiative and multivariate process analysis, monitoring and control.* 2006. **384**(5): p. 1043-1048.

7. Gernaey, K.V., et al., *A perspective on PSE in pharmaceutical process development and innovation.* 2012. **42**: p. 15-29.

571484 Process Performance of a Novel Continuous Process Line for Drying of an Active Pharmaceutical Ingredient

Lately, there has been increased interest in linking primary and secondary manufacturing. Especially, the increasing pressure towards cost effectiveness and cost reduction during phases II and III of drug product development is driving this demand.

After continuous API synthesis the steps filtration, washing and drying are still done in batch processes. This is mainly due to the lack of available equipment for pharmaceutical production. Additionally, drying processes are challenging. During crystallization particles are tailored with much effort and therefore, these properties must be maintained during subsequent process steps. However, during drying attrition and agglomeration compete, leading to a change of particle size distribution (PSD) of the primary particles.

This talk will show the design of a continuous process route for drying, considering also the upstream washing step, which strongly influences the agglomeration process during drying. This process line is an option to close the gap between primary and secondary continuous manufacturing.

The process starts with a continuous washing and filtration step. In the final filtration step the suspension is concentrated to the maximum solid content, which can be transported to the dryer. A novel contact convective drying technology was developed, which allows achieving a final moisture below 1% while not impacting the PSD. The process line is designed to dry cohesive temperature sensitive materials with a mass flow of 0.5 to 2 kg/h of dry substance. With this novel drying process, it is possible to gently dry particles with a particle size below 100 μm without changing the particle size. Also, this drying process guarantees that there is no agglomeration inside the dryer, ensuring a suitable residence time distribution. For this work Ibuprofen and other substances have been used to test PSD and residence time distribution, as well as drying efficiency. To prove that particle shape and surface did not change during the drying process, scanning electron microscopy was used.

In summary, it was shown that this novel contact convective drying technology is suitable to dry temperature sensitive cohesive particles at a mass flow range suitable for pharmaceutical production.

565447 On-Demand Drug Product Manufacturing in a Miniaturized System: Can Formulation be Simplified?

Oral direct compressible tablets are the most commonly used drug products. Tablets manufacturing requires design and development of formulations, which need to use several excipients. Choice of excipients depends on the concentration, manufacturability, stability, and bioavailability of the drug substance. Due to a large number of excipients typically used, formulation development becomes complex and tedious. At MIT, we developed a miniature platform for on-demand manufacturing of direct compressible tablets [1, 2]. On-demand manufacturing offers a great opportunity to simplify pharmaceutical formulations by minimizing the number of excipients required. In this study, direct compressible tablets of five pharmaceutical drugs, Diazepam, Diphenhydramine HCl, Doxycycline Monohydrate, Ibuprofen, and Ciprofloxacin HCl, with different drug loadings, were demonstrated. This miniature system is able to manufacture on-demand tablets from drug crystals by using the minimum number of excipients required for drug product performance. The critical quality attributes (CQA) of the tablets were assessed to meet the United States Pharmacopeia (USP) quality standards. All drug tablets met USP quality standards after manufacturing and after 2 weeks of accelerated stability test, except for slightly lower drug release for Ibuprofen. We foresee that advancing on-demand ready-to-use pharmaceuticals manufacturing with this type of flexible, miniature, emerging automated (integrated) system will simplify pharmaceutical formulation. This development route will provide significant economic benefits by reducing time-to-market and increasing product quality. In conclusion, this study confirmed that on-demand manufacturing can be used as a promising platform for simplifying the formulations design compared to traditional formulations.

References:

1. Azad M, Osorio Caicedo J, Brancazio D, Hammersmith G, Klee D, Rapp K, Myerson A 2018. A compact, portable, re-configurable, and automated system for on-demand pharmaceutical tablet manufacturing. Int. J. Pharm. 539 (1-2):157-164.
2. Myerson A, Azad M, Hammersmith G, Brancazio D, Osorio Caicedo J. Systems and methods for the fabrication of Ttablets, including pharmaceutical tablets. U.S. Provisional Patent Application US 2018/0271791 A1, Published on September 27, 2018.

568328 Slow Freezing Process Design for Human Induced Pluripotent Stem Cells By Modeling Intracontainer Variation

Human induced pluripotent stem (hiPS) cells are the emerging source of regenerative medicine products. Along with the successful clinical studies for treating diseases such as Parkinson's disease, or spinal cord injuries, the demand of hiPS cells is increasing. This trend indicates the necessity for establishing cryopreservation processes of hiPS cells for storage and transportation. Generally, there are two methods for freezing cells: slow freezing and vitrification. For hiPS cells, it is planed that the liquid containing hiPS cells are filled in to a container such as vial, cryopreserved, and then thawed at the point of use. Here, slow freezing, as compared to vitrification, can benefit from scalability, simplicity in operation, and no direct contact with freezing agent, such as liquid nitrogen. However, the process could suffer from the quality variation inside the vial (i.e., intracontainer), between vials (i.e., intercontainer), and between and among freezers (i.e., intra- and interfreezer).

We present a model-based design of slow freezing processes for hiPS cells considering the intracontainer quality variation. A single-cell model was developed by integrating models describing a radial and temporal temperature profile, cell volume change through transmembrane water transport, and intracellular ice formation during slow freezing. Given the cooling rate of the freezer, the vial diameter and material, and the cryoprotective agent type, the model can produce the maxima of cell volume change and the intracellular ice crystal volume as the cell quality indicators, and the required freezing time as the productivity indicator.

The developed model showed sufficient performance compared with a recently reported experimental study on hiPS cells. Upon this confirmation, the model was applied for three design cases. The choice of a cryoprotective agent, cooling rate, and vial diameter affected the quality evaluation more than the choice of the vial material. When considering the productivity as the design objective, the optimal vial diameter changed depending on the constraints of cell demand and the acceptable intracellular ice crystal volume. In the ongoing work, we are performing rigorous performance assessment of the model by extensive experiments. Advanced process design and optimization, e.g., dynamic temperature optimization, integrated design of freeze/thaw process, is another ongoing subject.

575056 Developing a Systematic Approach for the Collaboration between Process Systems Engineering and Catalysis to Accelerate the Improvements of Chemical Processes

A considerable number of chemical processes nowadays involve some sort of catalytic steps, where catalysts are utilized to convert reactants into intermediates or final products. In some cases, catalysts play a pivotal role in the selection and design of a chemical process, which in turn could lead to significant changes in the economic performance of that process. Therefore, catalysts research and development is of a primary concern to the process systems engineering community, however, there is no systematic approach to identify areas at which catalysts can significantly improve the outcomes of a chemical process. This work is aimed at proposing a systematic approach where catalysis and process systems engineering scientific communities can join efforts in order to accelerate the development of chemical processes. To illustrate an application of the proposed approach, a methanol from shale gas process is taken as a case study. While this process is well-established, the conventional catalyst ($Cu/ZnO/Al_2O_3$) used for the conversion of syngas to methanol takes place under high operating pressure and have low single pass conversion, leading to higher costs associated with the process. A techno-economic assessment is conducted to identify the most influential characteristics of a future methanol synthesis catalyst that could significantly enhance the process's economic performance.

571566 Novel LNG Boil-Off Gas Re-Liquefaction System Design Using Dual-Mixed-Refrigerant Cycle and an Ejector

During transportation of LNG using ships, the fuel is stored cryogenically at 111K and ambient pressure. Due to heat transfer from the environment to the tanks a significant amount of LNG (approximately 0.15% per day) evaporates, known as boil off gas (BOG). This may lead safety issues due the increase in tank pressure, or economic losses if the excess vented without being recovered. However, many methods have been considered, such as combustion, using it as fuel or venting to atmosphere but are undesirable either because of concerns over pollution, low efficiency or high operating cost.

Conceptually for the re-liquefaction system of the BOG is attractive, and have been a number of studies, there are two limitations. In most recent proposed re-liquefaction system, non-liquefied gas remains in the stream returned to the tank, with the surplus gas returned to the tank and still has low efficiency (approximately 41.3%). Thus it was proposed to combine two existing technologies, new dual mixed refrigerants cycle based on cascade cycles and a re-liquefaction system using ejector, and develop it by modifying design. In a novel process design that avoids the return of gas to the tank, while reducing the energy consumption of the compressor stage, to improve energy efficiency. A mathematical model is also introduced for more accurately calculating conditions at the ejector-outlet.

The results show, the power consumption of the compressor stage for BOG entering the system is reduced and performance indicators such as energy efficiency and coefficient of performance, related to operating cost, is much improved than earlier proposals. This study would be the basis of the development of the midstream industry in the aspect of energy consumption and efficiency of the LNG Boil-off gas re-liquefaction system.

578530 Crystallization Process Understanding and Optimization Using FBRM and Raman As PAT Tools

The presentation features the application of process analytical technology (PAT) for understanding the crystallization process in process chemistry development and scale-up. PAT tools such as FBRM, PVM and Raman can be applied as efficient process-monitoring tools to provide real time profile and reliable information much faster than off-line testing. Two challenging case studies will be presented as examples; where PAT tools helped us understand crystallization mechanisms, such as seeding point, nucleation, crystal growth and polymorphic form. This understanding enabled us improve filtration time on scale up by controlling the particle size distribution (PSD) and also optimizing solvent selection for correct polymorphic form. Such information is important for designing science-based process control strategies.

576388 Design and Synthesis of RNA Thermoregulators to Optimize Production of Essential Metabolites and Therapeutic Molecules in Bioprocesses

Bioprocesses nowadays are used to manufacture a wide variety of chemical compounds including polymers, biofuels, and therapeutic molecules. In order to increase the yield of these valuable metabolites, strategies like overexpression and deletion of the involved genes have been implemented. Nevertheless, some of these genes have been found to be responsible for bacterial growth and other essential cell functions, therefore the deletion of these genes would result in more desired product yield per cell but a decreased cell density, and in many cases the result will be cell death. Using genetic switches can be an alternative approach to keep the expression of the involved genes controlled. In this way, the production of the interest metabolite or bacterial growth could be regulated by the metabolic flux. One interesting approach that has been explored in this work is the use of RNA thermoregulators, which are translational controllers of gene expression that respond to temperature shifts by changing their conformational structure. At low temperature, they acquire a secondary structure that precludes the binding of ribosomes to the Shine-Dalgarno (SD) sequence; preventing mRNA translation. Moreover, when temperature increases, RNA thermoregulators reconfigure their conformation, allowing the binding of ribosomes to the SD sequence—which in turn initiates mRNA translation.

Although naturally-occurring RNA thermoregulators are associated with diverse biological processes (e.g. bacterial pathogenesis and life cycle of some bacteriophages), many synthetic analogs have been designed and characterized, for optimal control of gene expression in vivo. In this work, we designed and synthesized different families of synthetic RNA thermoregulators that serve as genetic switches to control different pathways, in Escherichia coli, involved in the production of essential metabolites and molecules with therapeutic properties. This work describes a novel platform technology capable of controlling metabolic flux via RNA thermoregulation in vivo for the production of great value metabolites.

563993 Techno-Economic Analysis of Emerging Vaccine Platform and Production Technologies: How Can We Produce Vaccines Faster and Cheaper?

Conventional vaccine production technologies cannot produce vaccines against new diseases at the rates and volumes required to stop the spread of future epidemics. Additionally, infant and child vaccination demands are often unmet in developing countries due to high upfront vaccine manufacturing facility capital investment costs, inflexibility of manufacturing facilities to produce a wide range of products, and due to cold chain logistic complexities.

Emerging vaccine platform technologies together with advanced bioproduction technologies can empower the production of a wide range of vaccine types, rapidly and at large volumes. Therefore, these exciting new technologies can address both rapid outbreak-response and routine mass vaccination challenges.

However, the vaccine production processes using these new and transformative technologies are not yet established. Furthermore, the total costs of vaccines obtained using these new platform technologies and advanced bioproduction modalities are not well understood. Hence, here process simulation results and the techno-economic analyses of the following 4 new promising vaccine platform technologies is presented: (1) RNA vaccines, (2) Generalized Modules for Membrane Antigen (GMMA), (3) ADDomer vaccines, and (4) humanized yeast-produced vaccines. Additionally, the production costs of these vaccine platform technologies both in conventional and advanced bioproduction facilities are evaluated. For the advanced bioproduction scenario, intensified processes are considered, whereby the facility-related capital and operating expenses can be reduced by containing the production processes inside laminar flow isolators.

Our work indicates that emerging vaccine platform technologies coupled with advanced bioproduction facilities can enable: (a) cost-effective, small-scale manufacturing, promoting out-scalable and distributed manufacturing; (b) rapid and on-demand production; (c) thermostable vaccine products, potentially eliminating cold chain requirements; and (d) flexible production of a wide range of vaccine products using the same production line.

569962 Synthesis of Optimal Hybrid Separation Processes

While distillation operations are the primary vehicles for separation within the chemical industry, they incur high utility costs due to their low thermal efficiencies [1]. Over 50% of energy required for distillation is used for purifying the last 5–10% of the distillate product [2]. This results in significant burden on the economics as well as on the environmental footprint of the overall operation when high-purity end products are targeted. Alternative techniques with membrane and adsorption-based separation technologies require the development of highly selective materials amenable to cost-effective scale up to compete with conventional standalone distillation equipment. Hybrid separation systems combining aspects of both technologies can be optimally designed to reduce energy input and costs while retaining advantages of conventional distillation units. Design of such hybrid systems requires process systems engineering approaches to identify the optimal topology of the separation sequence along with the operational conditions where the synergy is maximum. Determining the optimal topology, e.g. feed location for the distillation column, sequence and connectivity of the different operations involved, etc., in a sequential manner with scenario-based approaches may result in either combinatorial explosion or omission of potential favorable alternatives in an effort to restrict to a tractable number of possible scenarios [3]. Recently, the building block superstructure [4] representation has been demonstrated to provide a systematic methodology for process synthesis and intensification [5-6].

In this work, we extend the use of the building block representation approach for the design and synthesis of hybrid distillation-based separation schemes incorporating membrane and adsorption technologies. For a given design problem, the proposed approach starts with identification of the separation tasks that need to be accomplished. These problems can belong to either completely new process designs or part of an existing flowsheet. Identification of the separation tasks also helps into define targets for improvement. One major target for the hybrid scheme is that all the incorporated separation techniques operate at their highest efficiencies and none of them alone can do the same task better than the hybrid one. After the design targets are set, building block-based superstructure is formed for a hybrid scheme involving distillation, membrane, adsorption, etc. and solved as a synthesis-design-intensification problem addressing the identified separation task and design targets with the optimal design and operating conditions. This is facilitated by solving a mixed integer nonlinear programming (MINLP) problem describing the building block superstructure. To demonstrate the applicability of the proposed approach, optimal membrane-based schemes are synthesized for several hybrid separation schemes intended for solvent recovery applications. An adsorption-based hybrid distillation scheme is also designed for a cryogenic air separation to reduce the high energy input required especially for achieving >95% purities for non-hybrid distillation schemes. The resultant optimal solutions featuring hybrid schemes are examples of intensification that create special hybrid modules for use in processing routes where similar separation tasks exist.

References:

[1] Tian, Y., Demirel, S. E., Hasan, M. M. F., Pistikopoulos, E. N. (2018). An Overview of Process Systems Engineering Approaches for Process Intensification: State of the Art. Chemical Engineering and Processing: Process Intensification, 133, 160-210.

[2] Tula, A.K., Befort, B., Garg, N., Camarda, K.V. and Gani, R., (2017). Sustainable process design & analysis of hybrid separations. *Computers & Chemical Engineering*, *105*, pp.96-104.

[3] Marquardt, W., Kossack, S. and Kraemer, K., (2008). A framework for the systematic design of hybrid separation processes. *Chinese Journal of Chemical Engineering*, *16*(3), 333-342.

[4] Demirel, S. E., Li, J., and Hasan, M. M. F., (2017). Systematic Process Intensification using Building Blocks, *Computers and Chemical Engineering*, 105, 2-38.

[5] Li, J., Demirel, S.E. and Hasan, M.M.F., (2018). Process synthesis using block superstructure with automated flowsheet generation and optimization. *AIChE Journal*, 64(8), 3082-3100.

[6] Demirel, S. E., Li, J., and Hasan, M. M. F., (2019). A General Framework for Process Synthesis, Integration, and Intensification. Industrial & Engineering Chemistry Research. DOI: 10.1021/acs.iecr.8b05961.

569934 Process Intensification Framework for Extractive Separation Systems

The separation of azeotropic mixtures to achieve a higher purity beyond the azeotrope point is a frequent but challenging task in chemical processes. Extractive separation, a classic process intensification invention, offers the solution to overcome the physical equilibrium by introducing an extra mass separating agent (i.e., entrainer) [1-3]. Thus, the entrainers play a key role in these processes due to: (i) their solubility properties to enhance the system's mass transfer driving force to drive separation, and (ii) their hazardous properties (e.g., toxicity, flammability) for environmental and safety considerations.

Recent advances in the synthesis of process intensification systems have been leveraging phenomenological representation methods to systematically generate process solutions without pre-postulation of plausible equipment-based flowsheets which may hinder the discovery of novel design configurations [4-6], which also provides the opportunity to investigate the effect of functional materials at phenomenal level from process fundamental perspective [7]. However, the operational performances (e.g., flexibility, safety, controllability) in the resulting intensified designs, which are crucial to practical implementation, are mostly neglected at this synthesis stage [8]. Therefore, a holistic approach for the synthesis of extractive separation systems, with simultaneous consideration of entrainer selection and operability assessment, is still lacking.

To address this challenge, we have recently developed a systematic framework for the synthesis of operable process intensification systems [9], which features: (i) process synthesis, integration, and intensification using phenomena-based Generalized Modular Representation Framework (GMF) [10] to derive optimal design configurations, (ii) integrated steady-state operability and inherent safety considerations [11], and (iii) simultaneous design and control with explicit model-based predictive control strategies following the PAROC (PARametric Optimization and Control) framework [12]. In this work, we apply the proposed framework to synthesize enthanol/water extractive separation processes with guaranteed operability, safety, and controllability performances. Two entrainer candidates are considered and examined with respect to process profitability, safety, and environmental considerations: a conventional entrainer ethylene glycol (EG) and an ionic liquid entrainer 1-ethyl-3-methyl-imidazolium acetate ([EMIM][OAc]). Mass and/or heat integration are also simultaneously explored to reduce energy consumption, resource utilization, and waste production.

Reference

1. Mahdi, T., Ahmad, A., Nasef, M. M., & Ripin, A. (2015). State-of-the-art technologies for separation of azeotropic mixtures. *Separation & Purification Reviews*, *44*(4), 308-330.
2. Tian, Y., Demirel, S. E., Hasan, M. M. F., & Pistikopoulos, E. N (2018). An Overview of Process Systems Engineering Approaches for Process Intensification: State of the Art. *Chemical Engineering and Processing: Process Intensification*, 133, 160-210.
3. Zhou, T., Song, Z., Zhang, X., Gani, R., & Sundmacher, K. (2019). Optimal Solvent Design for Extractive Distillation Processes: A Multi-objective Optimization based Hierarchical Framework. *Industrial & Engineering Chemistry Research*. **DOI:** 10.1021/acs.iecr.8b04245.
4. Demirel, S. E., Li, J., & Hasan, M. F. (2017). Systematic process intensification using building blocks. *Computers & Chemical Engineering, 105*, 2-38.
5. da Cruz, F. E., & Manousiouthakis, V. I. (2016). Process intensification of reactive separator networks through the IDEAS conceptual framework. *Computers & Chemical Engineering, 105*, 39-55.
6. Tula, A. K., Babi, D. K., Bottlaender, J., Eden, M. R., & Gani, R. (2017). A computer-aided software-tool for sustainable process synthesis-intensification. *Computers & Chemical Engineering, 105*, 74-95.
7. Ismail, S. R., Pistikopoulos, E. N., & Papalexandri, K. P. (1999). Modular representation synthesis framework for homogeneous azeotropic separation. *AIChE journal*, *45*(8), 1701-1720.
8. Baldea, M. (2015). From process integration to process intensification. *Computers & Chemical*

Engineering, *81*, 104-114.

9. Tian, Y., Mannan, M. S., & Pistikopoulos, E. N. (2018). Towards a systematic framework for the synthesis of operable process intensification systems. In *Computer Aided Chemical Engineering*(Vol. 44, pp. 2383-2388). Elsevier.
10. Papalexandri, K. P., & Pistikopoulos, E. N. (1996). Generalized modular representation framework for process synthesis. *AIChE Journal*, *42*(4), 1010-1032.
11. Tian, Y., & Pistikopoulos, E. N. (2019). Synthesis of Operable Process Intensification Systems—Steady-State Design with Safety and Operability Considerations. *Industrial & Engineering Chemistry Research*. **DOI:** 10.1021/acs.iecr.8b04389.
12. Pistikopoulos, E. N., Diangelakis, N. A., Oberdieck, R., Papathanasiou, M. M., Nascu, I., & Sun, M. (2015). PAROC – An integrated framework and software platform for the optimisation and advanced model-based control of process systems. *Chemical Engineering Science*, *136*, 115-138.

566916 Process Intensification Via Batch-to-Continuous Transition in the Production of Lubricants

Production of large volume specialty chemicals to-date is conducted almost exclusively in large volume batch or semi-batch reactors. Transition to much smaller continuous processing technology offers the prospect of strong improvements over the century-old batch technology in terms of lower unit energy consumption, smaller footprint, less waste generation, increase in process reliability, and more consistent product quality. The present project, a close collaboration between the University of Pittsburgh and Lubrizol Corporation, aims to demonstrate this transition for succinimide dispersant chemistry as a test case to develop a systematic methodology for batch-to-continuous transition via rapid acquisition of reliable reaction kinetics, implementation in accurate reactor models, validation in a continuous tubular reactor, and, ultimately, construction and demonstration at pilot plant scale. In the present contribution, we focus on validation of reaction kinetics derived from batch experiments in a lab-scale continuous reactor, investigation of the major differences between the two processing technologies and highlight some of the key design considerations for the pilot plant continuous reactor.

The production of succinimide dispersants occurs via amination of a poly-isobutylene succinimide anhydride (PIBSA) with a polyamine. The reaction proceeds via a two-step mechanism: An irreversible amination to an amide acid intermediate, followed by a reversible dehydration reaction to produce water and the desired dispersant molecule. A complete family of dispersants is made by varying molecular weight of PIBSA, structure of the polyamines, and stoichiometric ratio of the two reactants. In the batch process, the water that is co-produced with the desired dispersant product is continuously removed from the head space at the relatively high reaction temperatures, rendering the dehydration step effectively irreversible. In contrast, the presence of water in the continuous reactor renders the dehydration step incomplete due to reversibility and identifies water as a key "contaminant" in this reaction system. However, we find that water removal from the product stream in a subsequent evaporation step results in spontaneous decomposition of the amide acid to the desired dispersant product, resulting in quantitative recovery of the desired product which is chemically and physically indistinguishable from the batch product and fulfills commercial specifications. The presentation will present and discuss these findings and their implications for design of an intensified continuous process.

563339 A Modified Mixed-Acid Catalytic System for Beckmann Rearrangement of Cyclohexanone Oxime in a Microreactor

e-Caprolactam (CPL), the precursor of nylon-6, is mainly produced through Beckmann rearrangement of cyclohexanone oxime catalyzed by oleum in industry. To improve the mixing performance and reaction selectivity of this conventional process, a mixed-acid catalytic system which contains trifluoroacetic acid (TFA) and oleum was firstly proposed and modified in a microchemical system. The effects of experimental parameters on the content of octahydrophenazine (OHP), which are SO_3 concentration, temperature, TFA to oleum ratio (T/A), oleum to oxime ratio (A/O), and TFA distribution in two fluids, were investigated in the microreactor. The content of OHP was reduced with the decrease of temperature and the increase of SO_3 concentration. T/A, and A/O. In addition, a proper TFA distribution is also beneficial for the reaction. A very low content (<1.0 ppm) of OHP was reached under the optimized conditions, which is accord with the requirement of superior-quality CPL.

567421 Fast Pyrolysis of Pine Biomass in a Gas Solid Vortex Reactor

The increased global concern on the dependence on fossil resources has pushed a development in the research of alternative feedstocks for the petrochemical industry. Thermochemical conversion of lignocellulosic biomass into bio-fuels and high valuable chemicals is one of the promising routes to decrease the fossil fuel dependence [1,2]. Fast pyrolysis is one of the most promising thermochemical routes that has been considered for bio-oil production on large scale due to the high bio-oil yields (up to 70 wt. %) [3,4], its simplicity and relatively low capital investment [5]. In spite of the advantages, fast pyrolysis also faces several challenges. The lack of a complete understanding of the pyrolysis reaction network and kinetics makes the tracking of each individual specie (from feed to product) difficult. Yet some simplified kinetic models match the experimental results closely [6]. Another challenge is the selection of the best reactor technology to provide accurate temperature control and fast entrainment and quenching of the pyrolysis vapors (of the order of ms. instead of s).

Nowadays, the fluidized bed reactor (FBR) is one of the preferred technologies for fast pyrolysis due to its ease in design and operation. However, FBRs suffer from operational limits, eg. the maximum fluidization gas flow, while mass and heat transfer rates are moderate. These limits are observed to decrease in a fluidized bed operating in a centrifugal field. The Gas-Solid Vortex Reactor (GSVR) sustains a rotating solids bed by continuous tangential injection of fluidizing gas through multiple small injection slots. GSVRs are known to offer advantages over conventional fluidized bed reactors in the gravitational field, eg. the bed is more densely packed and gas-solid slip velocities are significantly higher. The latter results in enhanced heat and mass transfer rates.

At the Laboratory for Chemical Technology (LCT, Ghent University), biomass fast pyrolysis is carried out in a GSVR demonstration unit [7]. The goal is to further optimize the bio-oil yield and to determine the bio-oil molecular composition using advanced analytical techniques (GCxGC). The present work focuses on experimentally exploring the bio-oil yield in the GSVR demonstration unit. Additionally, CFD simulations including reaction kinetics of this experimental GSVR are performed [8]. By comparing numerical and experimental results, the performance of the selected kinetic model will be validated. First the elemental and molecular composition of the feedstock, and the biomass particle aspect ratio are closely examined. Fast pyrolysis experiments with pine are performed to determine the yield of the main pyrolysis fractions (i.e., bio-oil, char and non-condensable gases). GSVR is compared with conventional FBRs in terms of their respective pyrolysis product fraction distribution. Next, after each experiment, the elemental and molecular composition of the bio-oil obtained from the GSVR is compared with that of bio-oil from FBR technology.

References

[1] Mohan, D., et al. (2006). "Pyrolysis of Wood/Biomass for Bio-oil: A Critical Review." Energy & Fuels 20(3): 848-889.

[2] Czernik, S. and A. V. Bridgwater (2004). "Overview of Applications of Biomass Fast Pyrolysis Oil." Energy & Fuels 18(2): 590-598.

[3] D. Shen, W. Jin, J. Hu, R. Xiao, K. Luo (2015). An overview on fast pyrolysis of the main constituents in lignocellulosic biomass to valued-added chemicals: Structures, pathways and interactions, Renewable Sustainable Energy Rev., 51 761-774.

[4] M.S. Mettler, D.G. Vlachos, P.J. Dauenhauer (2012). Top ten fundamental challenges of biomass pyrolysis for biofuels, Energy Environ. Sci, 5 7797-7809.

[5] Bridgwater, A. V. (2012). "Review of fast pyrolysis of biomass and product upgrading." Biomass and Bioenergy 38: 68-94.

[6] Ranzi, E., et al. (2008). "Chemical Kinetics of Biomass Pyrolysis." Energy & Fuels 22(6): 4292-4300

[7] González Quiroga, Arturo, Pieter Reyniers, Shekhar Kulkarni, Maria del Mar Torregrosa Galindo, Patrice Perreault, Geraldine Heynderickx, Kevin Van Geem, and Guy Marin. (2017). "Design and Cold Flow Testing of a Gas-Solid Vortex Reactor Demonstration Unit for Biomass Fast Pyrolysis." Chemical Engineering Journal 329: 198–210.

[8] Kulkarni, S., Vandewalle, L., González Quiroga, A., Perreault, P., Heynderickx, G., Van Geem, K., & Marin, G. (2018). Computational fluid dynamics-assisted process intensification study for biomass fast pyrolysis in a gas–solid vortex reactor. ENERGY & FUELS, 32(10), 10169–10183.

Process Intensification: A Novel Approach for Development of Smaller, Cleaner and Energy Efficient Technologies.

Sirshendu Guha, Engineers India Limited, New Delhi, India

Abstract

Process Intensification, in brief, can be defined as development processes which are smaller, cleaner, safer and highly energy efficient. Likewise any "multidimensional concept, process intensification is not easily defined[1]". Although experts have provided diverse definitions on process intensification "the common thread among these definitions is a focus on new schemes and equipment that create improved processes by combining, controlling and or enhancing the chemistry and transport phenomena in a chemical process[1]". As per European roadmap, one of the main objectives of process intensification is innovative process and equipment design which can ensure enhancement in efficiency in processes and reduce both capital and operating expenses[1]. Although "Process Intensification efforts have lagged in US for last forty years compared to other parts of the world, industry and academia in Europe have already embraced developments in Process Intensified Technologies[2] ". In this context, it is worth mentioning that an institute named as RAPID (Rapid Advancement in Process Intensification Deployment) has been set up in US to spearhead the research and development activities in Process Intensification technologies. In India, some process intensification works have been carried out. A notable example may be development of Micro-Fluidic device such as Micro-Coiled Flow Inverter (MCFI) which provides enhanced heat transfer characteristics[3]. Process Intensification may be considered as a revolutionary approach to process design, development and implementation as it has potential to give birth to better products and processes which are safer, cleaner, smaller and cheaper. It may also lead to manufacture of new products which could not be produced by conventional process technologies. Intensified processes can result in to reduction in size of plant or equipment by 10 to 100 times. Process Intensification, when implemented in plant level will lead to de-compartmentalization of chemical unit processes and will advocate their integration into one whole.

Fundamentally, Process Intensification of a chemical plant can be accomplished by enhancing the transfer co-efficient or the driving force for mass and heat transfer. The prime reason for large distillation and absorption towers is that the inter-phase mass transfer rates are governed by gravity. One way to enhance the transfer rates considerably is to replace the gravitational field with centrifugal field which is higher by few orders of magnitude. Applying this principle, compact size Spinning Disk Reactors, Rotating Packed Bed Separators and Centrifugal Absorbers have been developed. Conventionally, for gas separations, large and expensive cryogenic distillation systems are used. In order to achieve gas separations in compact size equipment, moving bed adsorption technology can be effectively used which ensures utilization of the entire bed due to counter current contact of adsorbent solid and the gases. The counter current contact leads to increase in mass transfer driving force by ensuring non-equilibrium between fluid and solid phases and due to high mass transfer rate, the size of the equipment can be reduced to a large extent. Since, absorbers and extraction columns sizes are governed by flooding velocities, use of porous medium for the solvent may be useful in effecting large throughput through a compact size column.

Process Intensification goals can also be achieved by combining reaction and separation and carrying out the same in one equipment if the reaction improves the separation substantially by means of enhanced mass transfer or the separation leads to higher reaction rate. Reactive Distillation, Reactive Absorption technologies are the examples in this regard.

Clathrate hydrates or gas hydrates are crystalline water based solids in which small non-polar gaseous molecules such as CH_4, H_2 , CO_2 etc., can be trapped inside the cages of the H- bonded water molecules mostly under cryogenic conditions or at high pressures. Clathrate hydrates are not chemical compounds as the gaseous molecules are not bonded to the lattice. Clathrate hydrates can be decomposed easily by either reducing the pressure or increasing the temperature. Thus large quantities of hydrocarbon gases (CH_4, C_2H_6, and C_3H_8) and H_2 might be stored as hydrates under appropriate conditions and can be used as solid fuels which can be decomposed by reducing pressure for combustion. Deposits of Clathrates of methane have been found to occur naturally (around 6.4 trillion tones) on deep ocean floor. Such deposits have been found on the Norwegian continental shelf in the northern headwall flank of the Storegga Slide. Use of these natural gas hydrates may prove to be economical in near future once cost effective technology for their extraction is developed.

Finally, Process Intensification, now a day, is a proven way for size reduction of process equipment and in turn the entire process plant can be miniaturized. It may eventually lead to lower CAPEX (capital expenditure) and OPEX (operating expenditure) due to substantial reduction in equipment sizes and enhancement in energy efficiencies. It also leads to inherently safer designs by virtue of creation of smaller equipment. In conclusion, development initiatives in process intensification field are extremely encouraging and miniaturization of process equipment and process plants is a reality today.

Introduction

Ramshaw[4] among others pioneered work in the field of 'Process Intensification' (PI) as early as in 1970. Ramshaw & Cross[5] defined Process Intensification as "Process Intensification is a term used to describe the strategy of reducing the size of a chemical plant needed to achieve a given production objective[5]". Further, Stankiewicz and Moulijin[6] stated that "any chemical engineering development that leads to a substantially smaller, cleaner and more energy efficient technology is process intensification[6]". Process Intensification aims at drastic improvements of performance of a process by rethinking the process as a whole. In particular it can lead to the manufacture of new products which could not be produced by conventional process technology. In short, Process Intensification is a revolutionary approach to process design, development and implementation. It results in better products and processes which are safer, cleaner, smaller and cheaper. Process Intensification, when implemented in plant level can lead to de-compartmentalization of chemical unit processes and will advocate their integration into one whole. The integrated processes such as Reactive-Distillation, Reactive absorption, Reactive-Filtration etc can be utilized in order to develop a Process Intensified chemical plant. For example, Eastman Chemicals had replaced their methyl acetate plant with a single reactive distillation column. Process Intensification also leads to higher process flexibility, improved inherent safety, enhanced energy efficiency and ability to use reactants at higher concentrations. Process Intensification can be divided into two broad categories namely equipment (hardware) and methods (software). Examples of hardware are Oscillatory Flow Reactors, Rotating Trickle Bed Reactor, Catalytic Plate Reactors, Polymer Film Compact Heat Exchangers etc. Examples of methodologies are Reactive Distillation, Reactive absorption, Supercritical fluid extraction, Solar energy etc.

Fundamentals of PI

Process Intensification of a chemical plant becomes fundamentally achievable if one can enhance the coefficients of heat and mass transfer for equipment employed in the chemical plant. It may be noted that the primary reason for large distillation and absorption towers is that gravity governs inter-phase mass transfer rates. Replacement of gravitational field with centrifugal field can raise transfer rates considerably as centrifugal field is higher by few orders of magnitude than the gravitational field. Centrifugal Absorbers, Spinning Disk Reactors and Rotating Packed Bed Separators were developed using this concept.

Expensive and large distillation systems which operate under cryogenic conditions are conventionally used in Gas Separation process units. Moving bed adsorption technology which ensures usage of the entire bed on account of counter current contact of gases and solid adsorbent can be utilized effectively for achieving separation of gases in compact equipment. Mass transfer driving force increases due to non-equilibrium between fluid and solid phases as a result of counter current contact. Size of equipment reduces to a large extent due to high mass transfer rate.

Since sizes of Absorbers and Extraction columns are determined by flooding velocities, usage of porous medium for the solvent can be extremely useful in sending high throughput through a compact size column.

One can also achieve goals of process intensification by carrying out separation and reaction in the same equipment in case the reaction enhances separation by means of higher mass transfer rate or the separation ensures higher rate of reaction. Reactive Distillation and Reactive Absorption technologies are examples which use this concept.

Some Research and Development Initiatives

Some of the interesting research developments achieved so far in the field of Process Intensification are described below:

Continuous Oscillatory Flow Reactor

Adam P Harvey, Malcolm R Mackley and Thomas Seliger[7] of University of Cambridge have developed a Continuous *Oscillatory Flow Reactor* for production of biodiesel. An Oscillatory

Flow Reactor is a continuous flow reactor which consists of tubes fitted with equally spaced orifice plate baffles. Oscillatory motion is superimposed on the net flow of process fluid maintaining plug flow regime. Oscillatory motion enhances heat and mass transfer rates which lead to reduction in residence time and this in turn reduces the size of the reactor. Trans-estarification reactions (Rxn #1) of rapeseed oil with methanol in presence of NaOH have been successfully carried out in Oscillatory Flow Reactor of 1.5 m length and 25 mm internal diameter to produce 'bio-diesel' (methyl or ethyl esters). The reaction is:

Triglyceride + 3 Methanol $^{[NaOH]}$ ↔ Glycerol + 3 Methyl Ester (bio-diesel) (Rxn #1)

The methyl ester molecules are 16-22 chain length and suitable for use in modern diesel engines. The schematic diagram of the reactor is shown in Fig-1.

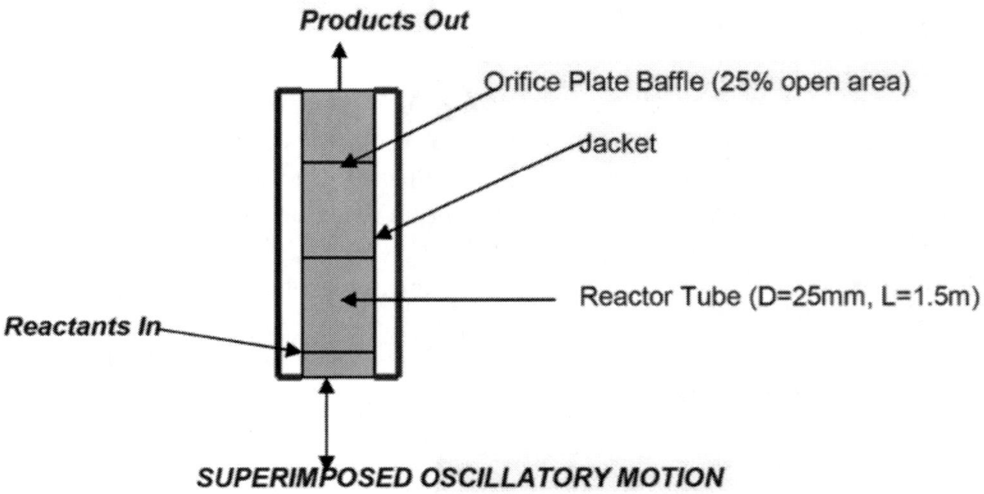

Fig-1: Schematic Diagram of an Oscillatory Flow Reactor

Catalytic Plate Reactor (CPR)

In Catalytic Plate Reactors (Fig-2), metal plates coated with suitable catalysts are arranged in such a manner that exothermic and endothermic reactions take place in alternate channels. These channels typically have a height of 1mm or so and the catalyst thickness is of the order of microns.

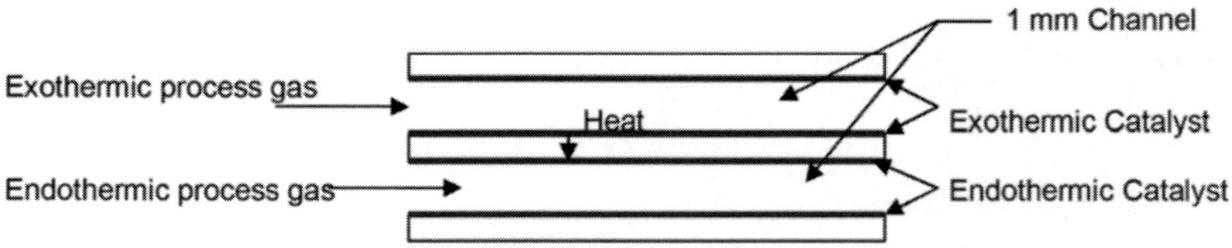

Fig-2: Schematic Diagram of a Catalytic Plate Reactor

In a CPR, heat transfer characteristics are excellent and intra-catalyst diffusion resistance is also minimal. The catalyst layers within the CPR are thin and this results in minimal diffusion limitations and hence catalyst utilization is very high. These factors leads to smaller reactor with lesser pressure drop. The size reduction factors with respect to current technology are as given in Table-1.

Table-1: Size reduction factors

Reaction	Conventional Technology	Size Reduction Factor
Steam Reforming	Packed Tubes in Furnace	More than 100
Fischer Tropsch Synthesis	Slurry Bed Reactor	4

Reactive Distillation (RD) Technology

In Reactive Distillation, reaction and separation are carried out in the same equipment in contrast to the conventional technique wherein reaction followed by separation scheme is followed. The notable example of RD Technology is replacement of Eastman Chemical Company's methyl acetate plant with a single Reactive Distillation column. The principal reaction for methyl acetate production is:

$$CH_3COOH + CH_3OH \leftrightarrow CH_3COOCH_3 + H_2O \qquad \text{(Rxn \#2)}$$

The conventional process for production of methyl acetate requires nine distillation columns whereas in RD Technology a single column with reactive, rectifying and stripping sections is adequate to produce methyl acetate of 95% purity. In conventional process, nine distillation columns are required because of formation of azeotropes between methyl acetate & methanol (Rxn #2), methyl acetate & water. In RD, nearly 100% conversion of acetic acid to methyl acetate is achieved due to continuous removal of products and furthermore, due to reverse reaction the azeotrope is broken with formation of four components mixture. Thus the reverse reaction aids in separation of components by breaking the azeotropes and separation of products leads to enhancement in conversion. This way, the RD Technology leads to process intensification and the CAPEX and OPEX are reduced to one-fifth of the conventional process[8]. In Fig-3, schematic diagram of a RD Column is shown.

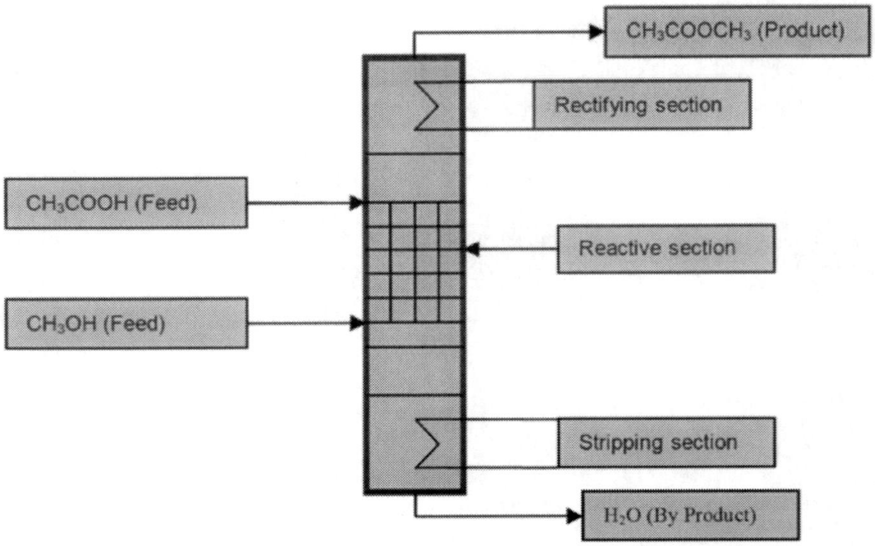

Fig-3: Schematic Diagram of a RD Column

Rotating Trickle Bed Reactor (RTBR)

In a Trickle Bed Reactor, mass transfer governs the reaction rates and Sivalingam et.al[9]., have shown that hydrogenation reaction rate of alpha methyl styrene over a rotating string of catalyst particles was about five to seven times as compared to the rate that could be achieved with gravity flow through string catalyst particles. Since the reaction rates can be substantially enhanced by replacing the gravitational field with centrifugal field, PI objectives could be achieved by development of Rotating Trickle Bed Reactors.

Future Development Idea

Clathrate Hydrate as solid fuel

Gases such as CH_4, H_2 , CO_2 etc., produce solid Clathrate Hydrates or Gas Hydrates in presence of water either at high pressures or at cryogenic temperatures. In these hydrates, the non-polar gaseous molecules of CH_4, H_2, CO_2 get trapped inside the cages of H-bonded water molecules under cryogenic condition or at high pressure. These hydrates cannot be considered as chemical compounds as the gaseous molecules of CH_4, H_2, CO_2 are not bonded to the lattice structure.

Clathrate hydrates can be sublimed easily either by increasing the temperature or reducing the pressure. Using this concept one can think of storing large quantities of light hydrocarbon gases such as CH_4, C_2H_6, C_3H_8 and H_2 under cryogenic condition or at high pressure in presence of moisture. These solidified light hydrocarbon gases can be used as solid fuels after suitably subliming into their respective gaseous states by reducing pressure or heating. Around 6.4 trillion tones of methane hydrates have been found to occur naturally on deep ocean floor on the Norwegian continental shelf in the northern headwall flank of the Storegga Slide. Once cost effective technology for extraction of these hydrates is developed, usage of them may be economical.

Conclusions

Process Intensification (PI) can be considered an established way for reduction of sizes of process equipment and hence miniaturization of an entire process plant. PI can also be understood as a smart and revolutionary approach to design of process equipment, development of processes and their implementation. PI has capability to give birth to better processes and products which are safer, cleaner, smaller and cheaper. PI leads to lower CAPEX and lower OPEX as a result of substantially lesser equipment sizes. It results into inherently safer design as a result of manufacture of smaller equipment. Finally, initiatives taken for development in PI field are highly encouraging and miniaturization of process equipment and process plants is a reality today.

Literature cited

[1] James Bielenberg and Michelle Bryner, CEP, March 2018, pp. 41-45

[2] T. Walsh, CEP, March 2018, pp. 40

[3] K.D.P. Nigam et al., Chemical Engineering and Processing, 86 (2014), pp.78-89

[4] Ramshaw, C.: *Chem. Eng. (London),* **389** (1983), 13

[5] Cross, W.T.; Ramshaw, C.: *Chem.Eng.Res.Des.*,**64** (1986), 293

[6] Stankiewicz, A.I.; Moulijin, J.A.: *Chem.Eng.Prog.*, **96** (2000), 22

[7] Adam P Harvey, Malcolm R Mackley and Thomas Seliger.: *Journal of Chemical Technology and Biotechnology*, **78 (2003)**, 338-341.

[8] Siirola, J.J., *AIChE Symposium Series*, **91** (304), 222-233 (1995).

[9] Sivalingam, G.,Radhika,M.,Rao,D.P., & Rao, M.S., *Industrial Engineering and Chemistry Research*, **41**(13), 3139-3144 (2002).

Cyber-security with Dynamic Watermarking for Process Control Systems

Jaewon Kim, Woo-Hyun Ko, and P. R. Kumar
Electrical & Computer Engineering, Texas A&M University, College Station, TX, U.S.A

Acknowledgment

This material is based upon work partially supported by
NSF Science & Technology Center Grant CCF-0939370,
The U.S. Army Research Office under Contract No. W911NF-18-10331,
The U.S. Army Research Laboratory under Contract No. W911NF-19-2-0033,
and U.S. ONR under Contract No. N00014-18-1-2048.

Abstract

While advances in information and communication technologies have resulted in greater flexibility, scalability, interoperability, and easier administration of process control systems, they have also enabled adversaries to exploit network or software vulnerabilities to subvert sensors or actuators in process control. Adversarial intruders can hack into the system through open networks and conduct malicious attacks to cause critical damage to the entire system. In this paper, we address the problem of cyber-security for process control systems. We specifically address the problem of attacks on sensor measurements. When incorrect sensor measurements are used in feedback control loops, it can lead to incorrect actuation that can cause instabilities or degrade performance.

We investigate a general-purpose solution called "Dynamic Watermarking (DW)" to detect tampering with sensor measurements. It is shown in **[9]** that this technique can be used to secure linear stochastic systems in the presence of malicious sensor nodes or compromised sensor measurements. In **[10]**, it has been shown that DW can detect attacks on autonomous vehicles.

The fundamental idea of the DW mechanism is that actuator nodes superimpose a small private excitation signal on the actuation commands, as shown in Fig. 3. This additive private excitation signal to the "Plant" causes a change in the measurements made Sensor Nodes, in comparison to what they would otherwise be. To detect if the sensor measurements reported are genuine, we design an "Attack Detector", as shown in Fig. 3, that examines if the reported sensor measurements are in conformity with the secret excitation signal. Specifically, it conducts two statistical watermark tests on the reported sensor measurements. It is proved in **[9]** that the Attack Detector can thereby detect malicious activity.

In this paper, we experimentally demonstrate that DW mechanism can be used to detect attacks on sensors in process control systems. We use a representative process control system, a system of two coupled water tanks, as shown in Fig. 1. It has two tanks and a basin. The tanks and the basin are cascaded vertically, so that the upper tank feeds the lower tank whose water flows down into the basin. The control objective is to maintain the water level of both the upper and lower tanks at specified set-points by controlling the voltage of the pump which drives the water from the basin up to the upper tank. The pressure sensors to the controller provide a the level of water in each tank as a feedback at the bottom of each tank.

We experimentally demonstrate the ability of DW by considering two different sensor attacks. The first is a "stealthy" attack. The attacker uses a model of the system dynamics to generate simulated measurements, and reports the simulated measurements rather than the real measurements. The second one is a "replay" attack, where an attacker stores a series of past measurements and replays them later. It is this attack that was reportedly used in the Stuxnet attack **[10]**.

First, we implement each attack and show that each leads to the failure of the water tank control system, causing overflows in the water tanks. Next, we employ DW throughput to secure the coupled water tanks system. It examines the veracity of the reported measurements by checking if they are appropriately correlated with the injected watermark signals. If the statistics of the measurements fail the tests of DW by exceeding predefined thresholds, then the attack detector concludes that the watermarks were distorted or removed, and raises an alarm. What recourse to take once an alarm has been raised generally depends on the particular system at hand. In this case, it stops the entire process for system safety, preventing system failure. We present experimental results showing that the DW mechanism indeed detects the malicious sensor attacks, and thus secures the process control system.

We thoroughly investigate the performance of DW, such as the detection time versus magnitude (i.e., the variance) of the watermark signal, and the parameters of the DW mechanism.

1 Introduction

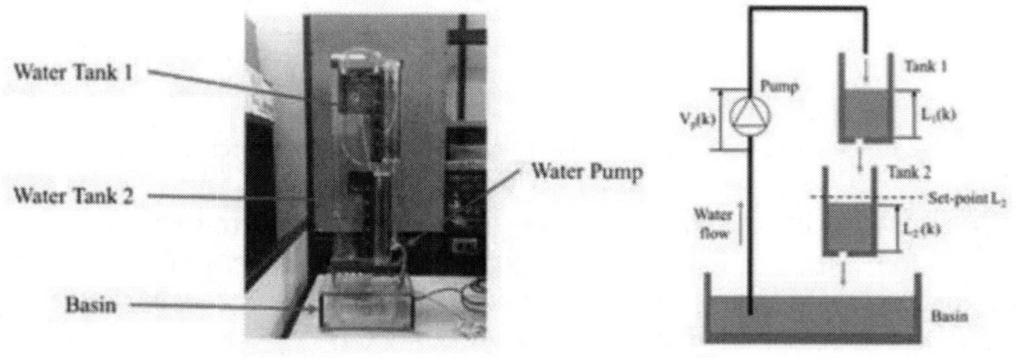

Figure 1: (a) The Coupled Water-tanks System (b) A Schematic Representation

The particular control system for the two tanks with two sensors and one manipulated variable is shown in Fig. 2a. The general representation of a linear stochastic control system shown in Fig 2b is convenient for describing the dynamic watermarking method. The **Nominal situation**, shown in Fig. 2b is that the honest sensor reports the real measurements $y(k+1)$ to the controller. We model the system as having two types of noises, the process noise $w(k)$, and actuator noise $w_u(k)$.

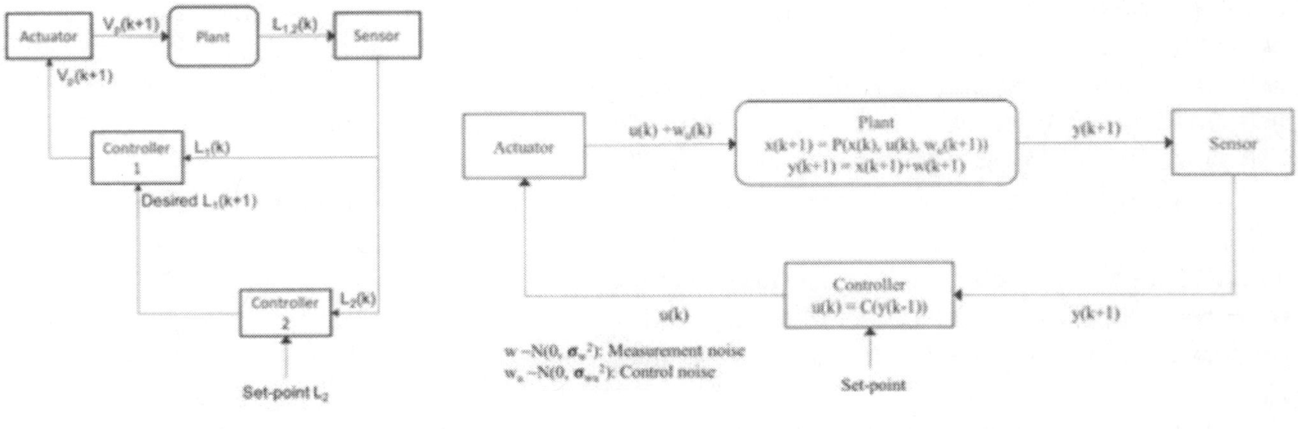

Figure 2: (a) Control System for Coupled Water-tanks (b) A General Linear Stochastic System Representation

2 Dynamic Watermarking (DW)

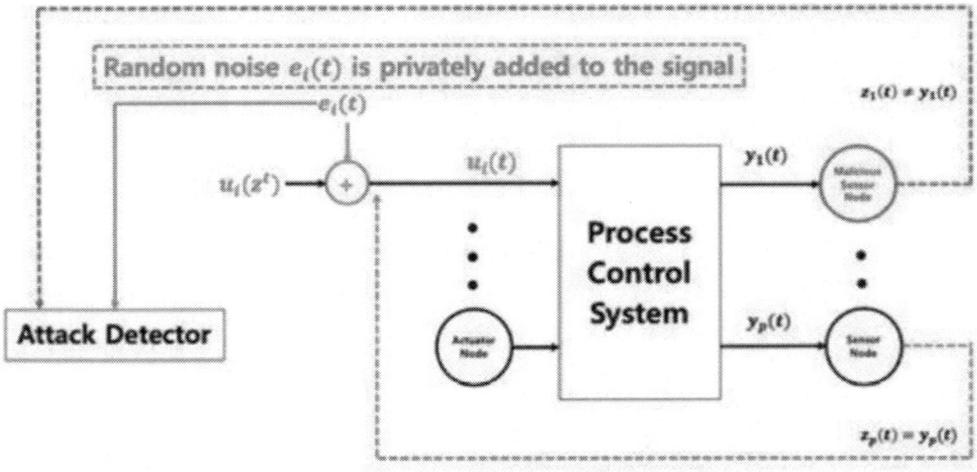

Figure 3: Dynamic Watermarking method for securing cyberphysical systems

The fundamental idea of DW is to have each actuator i add a random signal $e_i[t]$, known as the watermark, on the control policy-specified input $u_i(z^t)$. While the statistics of the watermark e_i are made known to every node in the system, its actual realization is not revealed to any other node $j \neq i$ in the system. Let independent and identically distributed $e[t] \sim \mathcal{N}(0, \sigma_e^2 I)$ denote the vector of watermarks added by all the actuator nodes. The system with output vector $y[t]$, manipulated variable vector u_t, and white gaussian noise $w[t]$ with zero mean and covariance matrix $\sigma_w^2 I$, is

$$y[t+1] = Ay[t] + Bu_t(z^t) + Be[t] + w[t+1]$$

Therefore, the sensor measurements y should satisfy the following equations if the sensor measurements are genuine:

$$(y[t+1] - Ay[t] - Bu_t(z^t)) \sim \mathcal{N}(0,\, BB^T\sigma_e^2 + \sigma_w^2 I)$$

$$E[e_i[t](y[t+1] - Ay[t] - Bu_t(z^t))] = B_i\sigma_e^2,$$

,where B_i = i-th column of B.

Based on this, each honest actuator $i \in 1, 2, ..., m$ subjects the reported sequence of sensor measurements $z(t)$ to the following two T:

TEST 1: The i-th node checks if the reported sequence of measurements satisfy

$$\lim_{T\to\infty} \frac{1}{T} \sum_{k=0}^{T-1} (z[k+1] - Az[k] - Bg_k(z^k))$$

$$(z[k+1] - Az[k] - Bg_k(z^k))^T = \sigma_e^2 BB^T + \sigma_w^2 I_n$$

TEST 2: The i-th node also checks if the reported sequence of measurements satisfy

$$\lim_{T\to\infty} \frac{1}{T} \sum_{k=0}^{T-1} e_i[k](z[k+1] - Az[k] - Bg_k(z^k)) = B_i\sigma_e^2.$$

In **[9]**, it is established that if the reported sequence of measurements passes both Tests, then any malicious sensor present could not have distorted the actual measurement value beyond adding a zero power signal to it.

Experimental Results for Several Types of Attacks

1. Stealth Attack: In this attack, beginning at time 70, the malicious sensor simulates the coupled water-tanks system with randomly generated actuator and measurement noises, and reports the resulting fictitious measurements $z(k+1)$ to the controller.

when watermark signal: mean(μ_e) = 0, var(σ_e^2) = 0.01

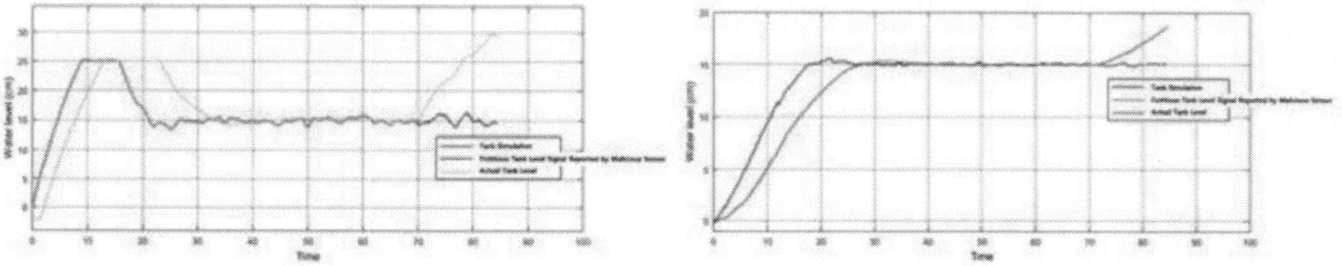

Figure 4: (a) Upper Tank (Tank 1)'s behavior (b) Bottom Tank (Tank 2)'s behavior

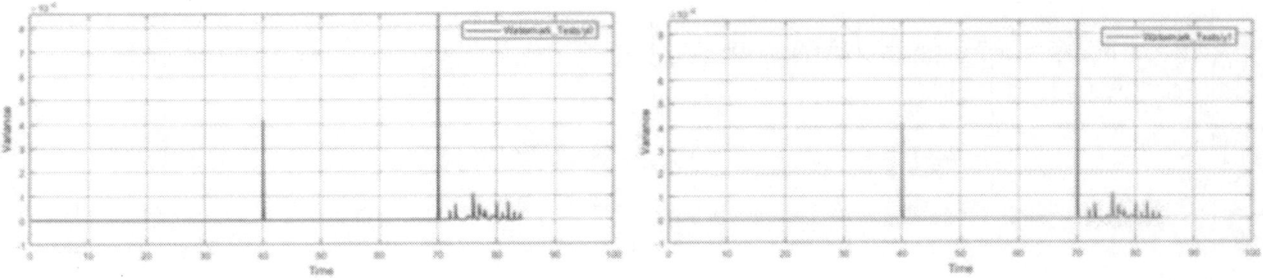

Figure 5: (a) Variance test 1 for Tank 2 (b) Variance test 2 for Tank 2

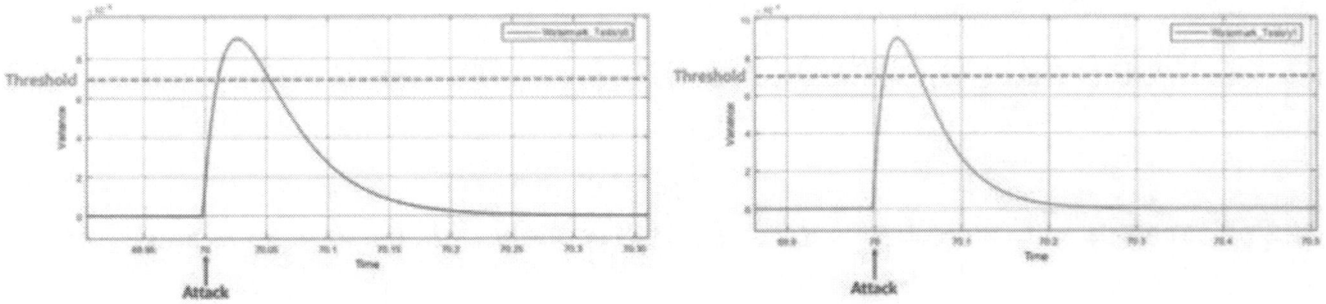

Figure 6: (a) Detection delay of Test 1 for Tank 2 (b) Detection delay of Test 2 for Tank 2

The two dynamic watermarking tests are started from 40 sec. As can be seen in the graph, the attacks result in the tanks are overflowing. However, as can be seen from the Figs. 6&7, the DW tests can detect less than 0.01 seconds after the attack begins. One may note that detection delay is very small even though the watermark signal is of very small variance, just 0.01, so that it does not affect the normal operation of the tanks.

2. Toggling Attack: In this attack, beginning at time 70, the malicious sensor flips the sign of actual tanks measurement values, and reports the resulting fictitious measurements $z(k+1)$ to the controller.

when watermark signal: mean(μ_e) = 0, var(σ_e^2) = 0.01

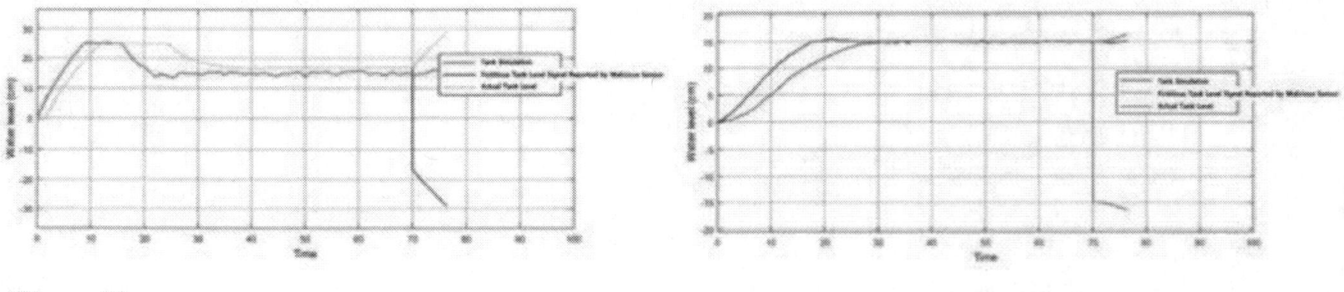

Figure 7: (a) Upper Tank (Tank 1)'s behavior (b) Bottom Tank (Tank 2)'s behavior

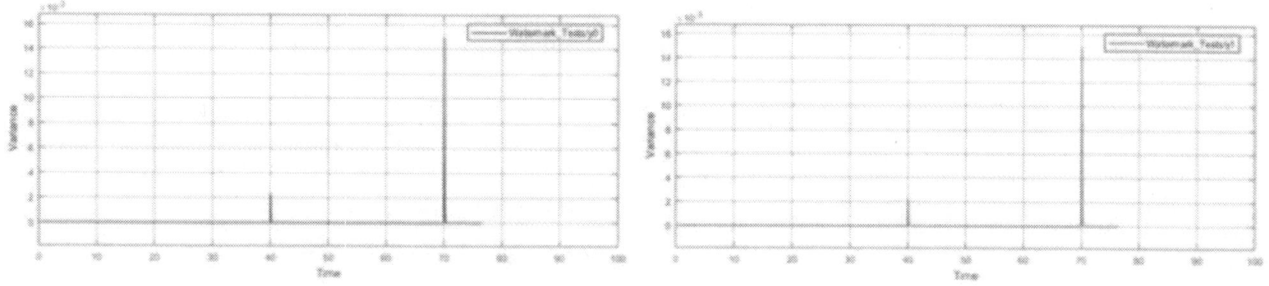

Figure 8: (a) Variance test 1 for Tank 2 (b) Variance test 2 for Tank 2

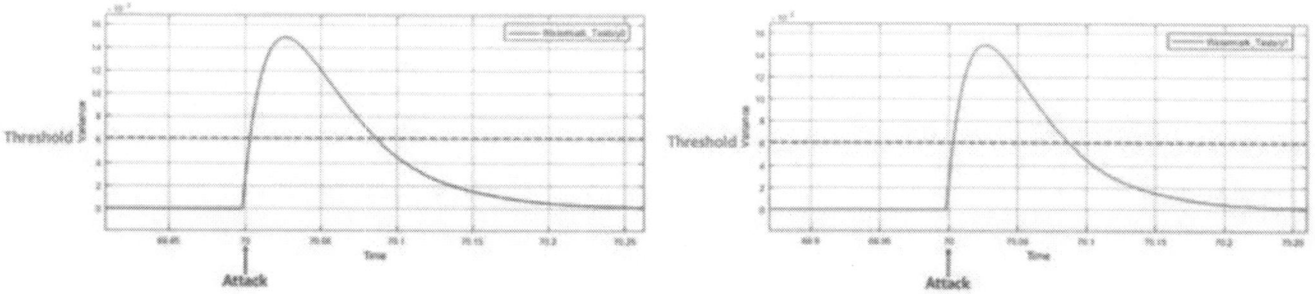

Figure 9: (a) Detection delay of Test 1 for Tank 2 (b) Detection delay of Test 2 for Tank 2

The two dynamic watermarking tests are started from 40 sec. As can be seen in the graph, the attacks result in the tanks are overflowing. However, as can be seen from the Figs. 8&9, the DW tests can detect less than 0.01 seconds after the attack begins. One may note that detection delay is very small even though the watermark signal is of very small variance, just 0.01, so that it does not affect the normal operation of the tanks.

3. Replay Attack: In this attack, from the beginning at time 40, the malicious sensor stores a series of actual measurements and replays them beginning at time 100, and reports the resulting fictitious measurements $z(k+1)$ to the controller.

when watermark signal: mean(μ_e) = 0, var(σ_e^2) = 0.01

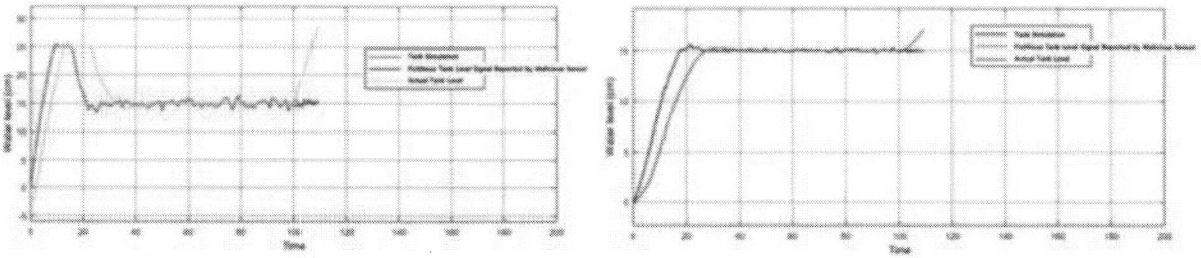

Figure 10: (a) Upper Tank (Tank 1)'s behavior (b) Bottom Tank (Tank 2)'s behavior

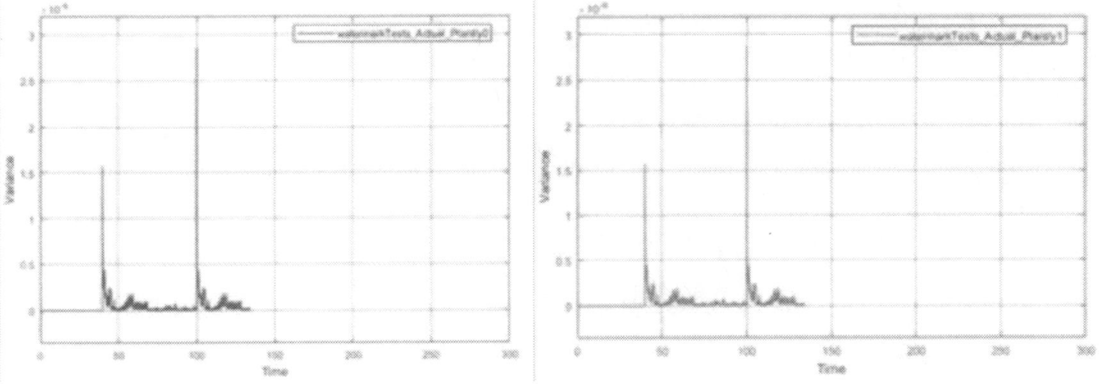

Figure 11: (a) Variance test 1 for Tank 2 (b) Variance test 2 for Tank 2

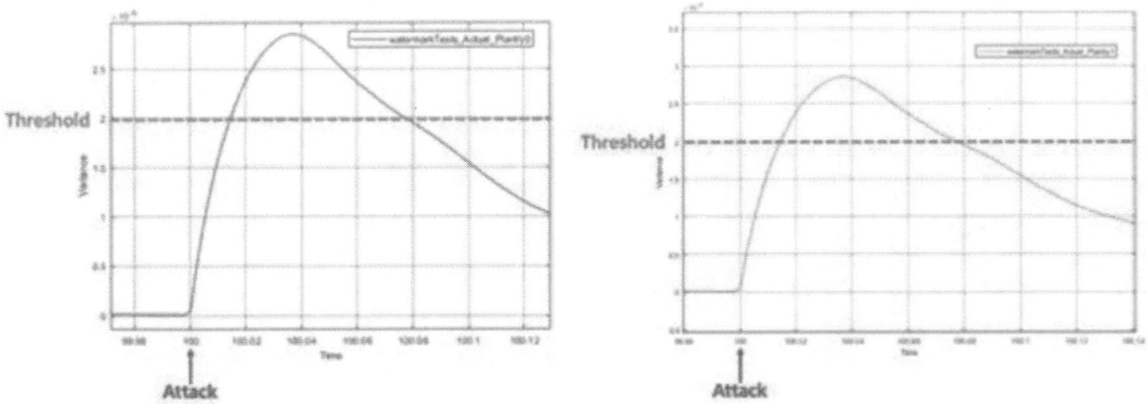

Figure 12: (a) Detection delay of Test 1 for Tank 2 (b) Detection delay of Test 2 for Tank 2

The two dynamic watermarking tests are started from 40 sec. As can be seen in the graph, the attacks result in the tanks are overflowing. However, as can be seen from the Figs. 11&12, the DW tests can detect less than 0.015 seconds after the attack begins. One may note that detection delay is very small even though the watermark signal is of very small variance, just 0.01, so that it does not affect the normal operation of the tanks.

Conclusion

This paper addresses the problem of cyber-security for process control systems. To illustrate the approach and results, we have presented experimental results on the Coupled Water-tanks System. The Dynamic Watermarking method is however a fairly general purpose method and can be used on other systems too. The general idea of dynamic watermarking is to superimpose private excitation signal on the actuation signals. Then by statistically checking whether the reported sensor measurement readings are compatible with the injected random watermark, any malicious tampering with the true sensor measurements can be detected. The advantage of the Dynamic Watermarking approach is that it secures the control system from arbitrary attack strategies. The Dynamic Watermarking method is backward compatible in that it can be implemented on legacy control systems by minimal modifications to the control logic. The effectiveness of Dynamic Watermarking for securing the coupled water-tanks system is indicative of its potential for securing more general process control systems.

References

[1] Ralph Langer. To kill a centrifuge. *The Langer Group, November*, 2013.

[2] Joseph Sangil Kwon, Bharadwaj Satchidanandan, Woo Hyun Ko, Jaewon Kim, Abhinav Narasingam, and P. R. Kumar. Securing process control systems using dynamic watermarking. In *Proceedings of 2018 AIChE (American Institute of Chemical Engineers) Annual Meeting (Computing and Systems Technology Division)*.

[3] Bharadwaj Satchidanandan and PR Kumar. Defending cyber-physical systems from sensor attacks. In *International Conference on Communication Systems and Networks*, pages 150–176. Springer, 2017.

[4] Woo Hyun Ko, Bharadwaj Satchidanandan, and P. R. Kumar. Dynamic watermarking-based defense of transportation cyber-physical systems. *ACM Transactions on Cyber-Physical Systems (TCPS): Special Issue on Transportation Cyber-Physical Systems.*

[5] Bharadwaj Satchidanandan and Panganamala Ramana Kumar. Secure control of networked cyber-physical systems. In *2016 IEEE 55th Conference on Decision and Control (CDC)*, pages 283–289. IEEE, 2016.

[6] Bharadwaj Satchidanandan and Panganamala Ramana Kumar. On minimal tests of sensor veracity for dynamic watermarking-based defense of cyber-physical systems. In *2017 9th International Conference on Communication Systems and Networks (COMSNETS)*, pages 23–30. IEEE, 2017.

[7] Bharadwaj Satchidanandan and PR Kumar. On the design of security-guaranteeing dynamic watermarks. *IEEE Control Systems Letters*, 4(2):307–312, 2019.

[8] Bharadwaj Satchidanandan and PR Kumar. Towards characterizing the watermark-securable subspace of a linear stochastic system. In *Proceedings of 2020 12th International Conference on Communication Systems and Networks (COMSNETS)*. IEEE, 2020.

[9] Bharadwaj Satchidanandan and Panganamala R Kumar. Dynamic watermarking: Active defense of networked cyber–physical systems. *Proceedings of the IEEE*, 105(2):219–240, 2016.

[10] Woo-Hyun Ko, Bharadwaj Satchidanandan, and PR Kumar. Theory and implementation of dynamic watermarking for cybersecurity of advanced transportation systems. In *2016 IEEE Conference on Communications and Network Security (CNS)*, pages 416–420. IEEE, 2016.

575039 Simultaneous Analysis of Control and Cyberattack Detection Algorithms for Enhanced Operational Safety

A critical component of the literature on cybersecurity of industrial control systems has been attack detection techniques. For state measurement falsification attacks, detection methods attempt to identify that an attack is occurring by noting that the measurement data does not match expected trends [1,2,3]. Because sensor measurement falsification cyberattacks make process inputs independent of the actual value of the process state [4], feedback controllers may not be able to prevent cyberattacks from causing harm unless they are notified, via a detection algorithm, that an attack has occurred and that a backup policy must be put in place. One of the drawbacks of detection strategies, however, is that they assume that an attack will take a form that is identifiable as an anomaly by the detection method.

Motivated by the above considerations, we explore the relationship between detection methods and optimization algorithms for chemical processes under model predictive control designs in the presence of state measurement cyberattacks. Specifically, we use a continuous stirred tank reactor example to demonstrate that when the state measurements remain close to the actual values (meaning that the attacks are less "anomalous"), the solutions of the optimization algorithms of the controllers may not differ significantly from the solutions when the state measurement is correct. However, as the state measurements differ more significantly (i.e., a potentially more detectable attack is pursued), the state trajectories can also differ more from the predicted trajectories within the controller. We therefore explore the intersection between the detection thresholds and the optimization problem solution's sensitivity to changes in the state measurement (i.e., initial condition of the process model) to better understand the conditions required for design of cyberattack-resilient detection-control/optimization algorithm combinations.

[1] Z. Wu, F. Albalawi, J. Zhang, Z. Zhang, H. Durand and P. D. Christofides, "Detecting and handling cyber-attacks in model predictive control of chemical processes," *Mathematics*, 6, 173, 22 pages, 2018.

[2] A. A. Cárdenas, S. Amin, Z.-S. Lin, Y.-L. Huang, C.-Y. Huang and S. Sastry, "Attacks against process control systems: Risk assessment, detection, and response," In *Proceedings of the ACM Asia Conference on Computer & Communications Security*, Hong Kong, China, 2011.

[3] I. Kiss, B. Genge, and P. Haller. "A clustering-based approach to detect cyber attacks in process control systems," In *Proceedings of the IEEE International Conference on Industrial Informatics*, Cambridge, UK, pp. 142-148, 2015.

[4] H. Durand, "A nonlinear systems framework for cyberattack prevention for chemical process control systems," *Mathematics*, 6, 169, 44 pages, 2018.

559885 Cybersecurity of Process Control Systems Via a Two-Tier Control Architecture

Since industrial control systems are usually integrated with numerous physical devices, the cybersecurity of control systems plays an important role in safe operation of industrial chemical processes. However, due to the use of a large number of control actuators and measurement sensors and the increasing use of wireless communication, control systems are becoming increasingly vulnerable to cyber-attacks, which is a series of computer actions aiming to compromise the security of control systems (e.g., integrity, stability and safety) and may cause severe industrial incidents [1]. Therefore, the design of advanced detection systems [2] and process control system for nonlinear processes in the presence of cyber-attacks remains an important issue.

In the context of improving process control using a large set of networked sensors and control actuators, a two-tier control architecture was developed for process control systems in [3] to improve plant performance accounting for high-frequency and low-frequency state measurements to handle the increasing number of sensors and actuators. Specifically, a lower-tier control system was developed using PI control based on wired, point-to-point communication with continuous measurements to ensure closed-loop stability, while an upper-tier control system was developed using model predictive control (MPC) based on networked (potentially wireless) communication with asynchronous measurements to improve closed-loop performance. However, in such a two-tier control architecture, the upper-tier control system is more vulnerable to cyber-attacks due to networked measurements. Motivated by this, a secure two-tier control architecture is developed to mitigate the impact of cyber-attacks targeting the upper-tier control system in chemical processes by integrating a machine learning-based detection method [4] with a two-tier control architecture. Specifically, a class of stealthy cyber-attacks that aim to compromise close-loop stability is first designed and applied to the upper-tier control system. Subsequently, a machine learning-based detection method is developed based on continuous state measurements from the lower-tier control system. After the cyber-attack is detected, the upper-tier control system is updated to mitigate the impact of cyber-attacks and re-optimize the closed-loop system. Finally, the effectiveness of the proposed secure two-tier control architecture is demonstrated through a chemical process network example.

[1] Ye N, Zhang Y, Borror C M. Robustness of the Markov-chain model for cyber-attack detection. IEEE Transactions on Reliability, 2004, 53: 116-123.

[2] Ozay M, Esnaola I, Vural F T Y, et al. Machine learning methods for attack detection in the smart grid. IEEE Transactions on Neural Networks and Learning Systems, 2016, 27: 1773-1786.

[3] Liu, J., Munoz de la Pena, D., Ohran, B. J., Christofides, P. D., and Davis, J. F. A two-tier control architecture for nonlinear process systems with continuous/asynchronous feedback. International Journal of Control, 2010, 83, 257-272.

[4] Wu, Z., Albalawi, F., Zhang, J., Zhang, Z., Durand, H., and Christofides, P. D. Detecting and Handling Cyber-attacks in Model Predictive Control of Chemical Processes. Mathematics, 2018, 6, 173.

574665 A Model-Based Approach for the Analysis and Mitigation of Cyber-Attacks in Networked Process Control Systems

With the significant growth in communication and networking capabilities in recent years, and the concomitant advances in sensor and actuator manufacturing technologies, process operations have become increasingly reliant on sensor, actuator and control systems that are accessed over real-time shared communication networks (wired and/or wireless) rather than dedicated, point-to-point links. The integration of a communication network in the sensor-controller and controller-actuator links gives rise to a networked control system and can substantially improve the flexibility and fault-tolerance capabilities of an industrial control system, in addition to reducing the installation, reconfiguration and maintenance time and costs. In addition to the economic and operational benefits of networked control architectures, networked control systems are an important enabling technology for realizing the vision of smart plant operations where advanced cyber-infrastructure and communication technologies are used to tightly integrate, in real-time, process control and operations with business decision making and sustainable environmental, health and safety performance [1].

The increased reliance on networked control systems, while beneficial from a plant performance point of view, poses a host of fundamental challenges due to the inherent limitations on the transmission and processing capabilities of the communication medium. Issues such as network resource constraints, real-time scheduling constraints, data losses and communication delays, may degrade the overall closed-loop performance and even lead to instability if not accounted for in the control system design. This realization has motivated significant research work on the analysis and design of networked control systems, and the literature on this topic is quite extensive (e.g., [2]–[4]).

Another key challenge with networked control systems, especially when dedicated local area control networks are integrated with wireless real-time sensor networks, is the cybersecurity challenge owing to the open nature of wireless networks. Recently, the security of control systems has become the focus of increasing attention due to the deployment of large sensor and actuator networks, as well as the increased use of wireless communication. With these advances, control systems are becoming increasingly vulnerable to cyberattacks, which are a series of computer actions that can compromise the stability and safety of control systems. Cyberattacks generally aim to perturb the process inputs and alter them from what they would have otherwise been under normal operation. Some of these attacks take the form of falsifying sensor measurements sent to the feedback controller, providing incorrect signals to the control actuators, or manipulating stored process data [5].

The communication network and cyber-infrastructure security challenge spans the entire global chemicals industry and petroleum production representing nearly $3 trillion in economic impact. In recent times, cyber-attackers have been successful in causing damage to a uranium enrichment plant, causing power outages, and contaminating a wastewater treatment plant (e.g., [6]-[7]). If left unchecked, these cyber-security risks can potentially lead to physical damage, injury, or death and therefore are a critical problem to address.

Typical approaches for handling control system cyber-security risks are based on computer science, information technology, computer hardware, or networking solutions (e.g., [5], [8]-[10]). More recently, efforts within the process control community have been initiated to address this problem from a control point of view, including the development of frameworks for detecting cyber-attacks and preventing their damage within the context of economic model predictive control [11]-[12].

An approach to reduce the risk of cyber-attacks in a networked control system is to minimize the control system's reliance on the network as much as possible. This idea has been pursued in earlier works (e.g., [13]-[14]) in the context of handling resource constraints. Specifically, a model-based networked plant-wide control structure that enforces closed-loop stability with minimal communication was developed. A set of predictive models were embedded within each local control system and, in conjunction with the local state measurements, the local control action was generated at times when communication between the plant subsystems was suspended, and the states of the models were updated when communication was permitted at discrete times. In doing so, a minimum communication rate that guarantees closed-loop stability could be determined.

While this approach is appealing in that it helps reduce the susceptibility of the control system to network-induced cyber-attacks, it does not provide robustness guarantees against random cyber-attacks. An assessment of the robustness of the networked control system to cyber-attacks is important to identify the fundamental limits within which the system can passively tolerate a cyber-attack, and through this assessment one can also identify the key parameters that can be used to actively mitigate the effects of these attacks when they arise.

Motivated by these considerations, the objective of this contribution is twofold. The first is to develop a framework for the design and analysis of model-based networked process control systems that have well-characterized robustness margins in the presence of certain types of cyber-attacks, and the second is to devise active strategies for mitigating the impact of such attacks. To meet these objectives, we initially construct a model-based plant-wide networked control system in which the local control systems communicate over a shared communication medium. The communication strategy aims to reduce network utilization by embedding predictive models within each local controller and updating the model states by exchanging state measurements over the network periodically at a certain minimum rate.

We consider cyber-attacks in the form of falsified sensor measurements and analyze the impact of using falsified measurements to perform the model state updates at the communication times. By modeling the cyber-attacks explicitly in the closed-loop system formulation, the stability of the networked closed-loop system can be assessed, and an explicit characterization of the stability region in terms of the attack-induced measurement errors, the communication rate and the controller design parameters can be obtained. This characterization then yields the feasible operating range for each system parameter within which robust stability is guaranteed under cyber-attacks, and also reveals the key parameters that can be adjusted in order to mitigate such attacks. Active mitigation strategies are then devised on the basis of the stability region characterization. Finally, the analysis and design results are illustrated using a representative chemical process example.

References:

[1] Christofides PD, Davis JF, El-Farra NH, Clark D, Harris KRD, Gipson JN. Smart Plant Operations: Vision, Progress and Challenges. *AIChE Journal*. 2007; 53(11):2734–2741.

[2] Hespanha JP, Naghshtabrizi P, Xu Y. A survey of recent results in networked control systems. *Proceedings of the IEEE*. 2007; 95:138–162.

[3] Christofides PD, Liu J, Munoz de la Pena D. *Networked and Distributed Predictive Control: Methods and Nonlinear Process Network Applications*. London: Springer-Verlag, 2011.

[4] You KY, Xie LH. Survey of recent progress in networked control systems. *Acta Automatica Sinica*. 2013; 39:101–117.

[5] Khorrami F, Krishnamurthy P, Karri R. Cybersecurity for Control Systems: A Process-Aware Perspective. *IEEE Design and Test*. 2016; 33:75–83.

[6] Langner R. Stuxnet: Dissecting a Cyberwarfare Weapon. *IEEE Security and Privacy*. 2011; 9:49–51.

[7] Clark RM, Panguluri S, Nelson TD, Wyman RP. Protecting drinking water utilities from cyberthreats. *Journal of American Water Works Association*. 2017; 109:50–58.

[8] Pang ZH, Liu GP. Design and Implementation of Secure Networked Predictive Control Systems under Deception Attacks. *IEEE Transactions on Control Systems Technology*. 2011; 20:1334–1342.

[9] Ten CW, Liu CC, Manimaran G. Vulnerability Assessment of Cybersecurity for SCADA Systems. *IEEE Transactions on Power Systems*. 2008; 23:1836–1846.

[10] Linda O, Manic M, McQueen M. Improving Control System Cyber-State Awareness using Known Secure Sensor Measurements. In: *Critical Information Infrastructures Security*, edited by Hämmerli BM, Kalstad Svendsen N, Lopez J. Berlin, Heidelberg: Springer Berlin Heidelberg. 2013; pp. 46–58.

[11] Durand H. A Nonlinear Systems Framework for Cyberattack Prevention for Chemical Process Control Systems. *Mathematics*. 2018; 6(9).URL http://www.mdpi.com/2227-7390/6/9/169

[12] Wu Z, Albalawi F, Zhang J, Zhang Z, Durand H, Christofides PD. Detecting and Handling Cyber-Attacks in Model Predictive Control of Chemical Processes. *Mathematics*. 2018; 6(10). URL http://www.mdpi.com/2227-7390/6/10/173

[13] Sun Y, El-Farra NH. Quasi-Decentralized Model-Based Networked Control of Process

Systems. *Computers & Chemical Engineering*. 2008; 32(9):2016–2029.

[14] Sun Y, El-Farra NH. Resource-Aware Quasi-Decentralized Control of Networked Process Systems over Wireless Sensor Networks. *Chemical Engineering Science*. 2012; 69:93-106.

574448 On the Intersection of Process/Equipment Design and Control System Cybersecurity

Cybersecurity of process control systems is an emerging direction of research investigation in the chemical process control literature. Though a number of works have looked at ways of seeking to combat cyberattacks on industrial control systems (e.g., through detection methods and subsequent strategies which modify signals in a control loop to revert to those which are not compromised by an attack [1,2]), an open research topic in process control system cybersecurity is developing a theoretical framework in which they can be understood. Our recent work [3] demonstrated that for processes described by systems of nonlinear ordinary differential equations, the most general definition of a cyberattack is a condition in which any input within the input bounds can be applied to the process, regardless of the process state. Because the input does not necessarily have a relationship to the state under a cyberattack, these attacks may be difficult to handle with control designs without detection mechanisms that do not fail to identify the attacks before they become problematic.

Motivated by the above considerations, we take an alternative perspective to handling cybersecurity through process control, and instead investigate the manner in which process designs contribute to safety concerns under cyberattacks. Specifically, inspired by the fact that safety relief valves [4] are a mechanical means for preventing unexpected incidents by changing the process dynamics, we explore the manner in which process dynamics and input bounds are related to the extent to which an attack can succeed. We also explore the concept that equipment failure, even that of downstream equipment from the equipment which attacked actuators may directly impact, could be attempted to be compromised during a cyberattack. We explore how modeling of equipment behavior under dynamic process conditions (through equations which describe, for example, stress and strain) may aid in better understanding how cyberattacks can create problematic operating conditions.

[1] Z. Wu, F. Albalawi, J. Zhang, Z. Zhang, H. Durand and P. D. Christofides, "Detecting and Handling Cyber-Attacks in Model Predictive Control of Chemical Processes," *Mathematics*, 6, 173, 22 pages, 2018.

[2] A. A. Cárdenas, S. Amin, Z.-S. Lin, Y.-L. Huang, C.-Y. Huang and S. Sastry, "Attacks against process control systems: Risk assessment, detection, and response," In *Proceedings of the ACM Asia Conference on Computer & Communications Security*, Hong Kong, China, 2011.

[3] H. Durand, "A Nonlinear Systems Framework for Cyberattack Prevention for Chemical Process Control Systems," *Mathematics*, 6, 169, 44 pages, 2018.

[4] T. Marlin, "Operability in Process Design: Achieving Safe, Profitable, and Robust Process Operations." McMaster University, Ontario, Canada, 2012.

569200 Cyber-Physical Security of Advanced Automated Continuous Pharmaceutical Manufacturing Pilot-Plant

Currently, Industry 4.0 concepts are being applied to pharma industry to achieve Pharma 4.0 paradigm. It introduces cyber-physical systems that enable the computerization of pharmaceutical manufacturing. Advanced control, automation, advanced analytics, big data, cloud computing, and internet of things (IoT) are some of the terms commonly referred within Pharma 4.0. It has many advantages but also have bigger challenges on the cyber-physical security side. Over the years, there have been many expensive and notorious incidents in other industry that have prompted the necessity of improved cyber security system. Stuxnet, Duqu, Flame and Shamoon were malware specifically designed to target Industrial Control Systems (ICS) and cause physical damage to the equipment, and inhibit processes from executing normally. Today, process manufacturing in general and the pharmaceutical industry in particular are seeing increasing numbers of cyber security threats. To protect themselves, companies have to change the way they operate by addressing cyber security threats and designing appropriate defenses. Pharmaceutical companies have to recognize their high visibility and attractiveness as targets. They must work to understand cyber security concepts, identify weak points and implement appropriate security measures [1]. Therefore, the cyber security of pharmaceutical manufacturing is highly desired not only economical point of view but also for patient safety perspectives.

The continuous pharmaceutical manufacturing (CM) is one of the most suitable platforms to implement Pharma 4.0 concepts. There are several advantages but also different scientific challenges in implementation and protection of CM [2]. One of the current major challenges for the continuous manufacturing is to protect it from any cyber-physical attack since it has more exposure to system and outside word.

In this work, the cyber security aspects of continuous pharmaceutical manufacturing process have been systematically investigated. The CM pilot-plant is controlled via industry standard control platform. A systematic framework including the methods and tools have been developed for proactive identification and mitigation of potential cyber-physical attacks on CM. A Scalence X208 industrial Ethernet switch (Siemens) has been integrated with the plant so that the network traffic data can be collected. The cyber security relevant software tools such as Snap 7, Wireshark, and Tripwire have been integrated with the CM. The Snap 7 is used to monitor the data blocks to and from the PLC. Wireshark is used for packet sniffing purposes, and the Tripwire is used for the file integrity monitoring. The security data collected from these tools together with the process data collected through HMI (human machine interface) have been utilized to show the behaviors of the system both during normal operation as well as during a simulated attack. Additionally, this work was also focused on distinguishing the malicious attempts from common machine and physical failure.

The objective of this presentation is two folds. First to highlight the developed systematic cyber-security framework and second to demonstrate its application for the cyber-physical security of continuous pharmaceutical manufacturing pilot-plant.

References

1. Lee, C. H. (2018). Boosting Cybersecurity in Pharma. Pharma manufacturing magazine. https://www.pharmamanufacturing.com/articles/2018/boosting-cybersecurity-in-pharma/
2. Bhaskar, A., Barros, F. N., Singh, R. (2017). Development and implementation of an advanced model predictive control system into continuous pharmaceutical tablet compaction process. International Journal of Pharmaceutics, 534 (1-2), 159-178. https://doi.org/10.1016/j.ijpharm.2017.10.003.

567370 Simultaneous Product and Process Design for Tailor-Made Biofuels

By consuming roughly half of global oil produced [1], road transport contributes significantly to climate change. Internal combustion engines (ICEs) are currently the main propulsion system in road transport and are expected to stay important at least until 2050 [2]. Renewable fuels that are tailored for high-efficiency and low-emissions combustion pose a promising option for future ICEs. These fuels, which typically consist of multiple components, must be produced by economic and sustainable processes. In case the fuel components can be derived from the same feedstock in a single process, integrated design of process and product is needed to find the overall optimal process-product combination [3].

While many methods for process design and product design exist, few have dealt with simultaneous process and product design. Marvin et al. [4] propose a method for simultaneous fuel and reaction pathway design that is based on automatically constructed reaction networks. Following a similar approach, Dahmen and Marquardt [5] design tailor-made biofuels and their optimal production pathways using relatively detailed fuel property models that take into account non-ideal thermodynamics. The pathway models used by Marvin et al. [4] and Dahmen and Marquardt [5], however, neglect separations and their energy demands, thus hindering direct optimization of two key process performance measures, i.e., fuel production cost and global warming impact (GWI). By contrast, the optimization-based pathway screening method Process Network Flux Analysis (PNFA) proposed by Ulonska et al. [6] also evaluates downstream processing alternatives and their minimum energy demand based on reduced-order separation models [7]. PNFA requires a priori specification of a target product and therefore does not facilitate integrated product and process design; however, its more detailed mass- and energy-based pathway evaluation allows optimizing for fuel production cost and GWI.

We present a new method for integrated fuel and production process design that combines two of our aforementioned approaches, i.e., [5] and [6]. The new method is capable of performing simultaneous fuel and production process design with cost and GWI as objectives. To enhance computational tractability of the optimization problem, the new approach contains several model simplifications, e.g., an algebraic fuel distillation curve model based on the concept of the true boiling point curve [8]. These simplifications allow solving the resulting mixed-integer nonlinear program (MINLP) with the global deterministic solver BARON [9]. We demonstrate the new method with a case study on future tailor-made spark-ignition engine fuels consisting of up to seven potential bio-derived components. An analysis of the tradeoff between fuel production cost and GWI reveals how the composition and the process configurations of the rationally formulated fuels change along the Pareto-optimal curve from the minimal cost design to the minimal GWI design.

References:

[1] Statista. Distribution of oil demand worldwide in 2016 by sector. https://www.statista.com/statistics/307194/top-oil-consuming-sectors-worldwide/ (accessed March 2019).

[2] ERTRAC Working Group on Energy & Environment. Future light and heavy duty ICE powertrain technologies., 2016, https://www.ertrac.org/uploads/documentsearch/ id42/2016-06-09_Future%20ICE_Powertrain_Technologies_final.pdf (accessed March 2019).

[3] Hechinger M, Voll A, Marquardt W. Towards an integrated design of biofuels and their production pathways. Comput. Chem. Eng. 2010;34(12):1909–1918.

[4] Marvin WA, Rangarajan S, Daoutidis P. Automated generation and optimal selection of biofuel-gasoline blends and their synthesis routes. Energy Fuels. 2013;27(6):3585–3594.

[5] Dahmen M, Marquardt W. model-based formulation of biofuel blends by simultaneous product and pathway design. Energy Fuels. 2017;31(4):4096–4121.

[6] Ulonska K, Skiborowski M, Mitsos A, Viell J. Early-stage evaluation of biorefinery processing pathways using process network flux analysis. AIChE J. 2016;62(9):3096–3108.

[7] Bausa J, Watzdorf Rv, Marquardt W. Shortcut methods for nonideal multicomponent distillation: I. Simple columns. AIChE J. 1998;44(10):2181–2198.

[8] Reiter AM, Wallek T, Pfennig A, Zeymer M. Surrogate generation and evaluation for diesel fuel. Energy Fuels. 2015;29(7):4181–4192.

[9] Tawarmalani M, Sahinidis NV. A polyhedral branch-and-cut approach to global optimization. Math. Program. 2005;103(2):225–249.

573407 Improving the Location Robustness of Spatially Distributed Variables for Biomass Supply Chains Under Supply Uncertainty

Due to climate change and other issues surrounding fossil fuels, finding alternative fuels and feedstocks is of vital importance. By expanding the role of the bioeconomy, biomass can be utilized as a substitute for both uses. Supply chain optimisation (SCO) is a method that can been used to help the biomass industry gain a foothold and expand while remaining economically viable. One issues is while supply chain network optimization offers a dependable method of searching a large solution space to find results that are feasible and maximized (or minimized) for the objective function, it does have a weakness in that small differences in initial conditions can cascade and result in non-negligible variations in the form of the optimized result. Frequently the actual value of the objective is similar for small differences in initial conditions, but the organization of the network that is recommended by the optimization can be measurably different. Spatially distributed variables can be noticeably different for different starting conditions.

Spatially distributed variables are variables in the biomass supply chain optimisation which may be distributed over the real map space, such as locations to build biochemical plants, locations for feedstock storage, or locations to grow biomass. These variables are decided in the process of optimising the supply chain for the objective variable – whether to place a plant in a specific location, or how much biomass to grow in each location. Each potential organisation is a solution which will have different values of the objective variable. These spatially distributed variables can be sensitive to changes in input values, as the supply chain optimises for the new starting conditions. The new optimal solution recommends placing refineries, storage facilities and growth sites in different locations depending on the starting conditions. Ideally these variables would be locationally robust.

Location robustness is the idea that there should be some consistency in the final design of the optimised supply chain for minor variation in the inputs. Input variation is bound to happen, and ideally these supply chain designs should be able to perform to a high level regardless of the variation in input. For some applications this is not a significant concern; for high investment situations, this may be problematic. A final supply chain design can be very expensive and time-consuming to implement, and a design which is optimal for the given input but is sensitive to variations in that input would be less useful than designs which can perform across a range of conditions. When a company is deciding where to locate refineries, where to grow crops or where to store products, the investment demands that a location be viable regardless if the input prices or yield rates vary slightly.

One major source of input variation for these supply chain systems is from uncertainty in the supply of biomass, such as year-to-year variation in biomass yield, or in land quality drift due to environment changes. It is also one that is particularly difficult to control – weather is still a factor in the productivity of farming. Biomass yield is a key input for biomass supply chains, which can have cascading impacts on the final biomass supply chain design. Understanding how uncertainty from this input in particular can be mitigated is key to designing supply chain frameworks and supply chains which are locationally robust.

This work seeks to develop a biomass supply chain optimization framework that considers location robustness as part of the framework. The work first seeks to identify the causes of this locational instability within some supply chain frameworks to understand how this arises. It then develops techniques that may be more locationally consistent due to the construction of the framework.

Biomass supply chain models often use the mean biomass yield of cellularised tracts of land to calculate yield for that cell. However, there can be large variety in the biomass yield within those tracts, losing useful information to aggregation and simplification. A biomass SCO framework has been developed which can incorporate some of this information about the quality of land available by piecewise linearly approximation of the biomass yield distribution, and incorporating this information into the optimisation.

Using linear approximations of the biomass yield distributions allows the SCO model to make more accurate decisions about quantity and location of biomass growth operations, thereby affecting all downstream decisions as well. It is postulated that providing the optimisation with the option to focus on developing high yield land, that high quality land will be used more consistently. This would have the dual effects of being more reasonable from a development perspective – better to develop high quality land – and of potentially increasing the locational robustness. The good land will be used preferentially, reducing the number of new or different areas which may be developed for biomass growth under supply uncertainty.

A case study of the United Kingdom for potential biomass industry viability will be examined using the framework to illustrate the impact of retaining this biomass yield information in the optimisation. Initial results indicate that using the more accurate biomass yield estimates reduced the land overall usage. The framework results in a better use of the available land to meet available demands, reducing land use. Further, it allows for improved biomass output, which in turn increased the quantity of bioproducts which can be produced. All of this leads to an increase in the overall profit. The next step is to implement supply uncertainty and to examine the locational robustness of the framework with respect to these variations. This framework could be used with other optimisations with spatially distributed resources, where more detailed information could improve results, such as solar insolation or wind resource availability.

571188 Process Intensification in By-Product Recovery from Biobased Effluents

Low-value by-product processing is an economic challenge common to many biobased processes. For recovery, the downstream processing must be very efficient, Value-adding conversion via chemical reaction may be beneficial. Both goals can be met with reactive separations. Chemical conversion of target constituents may simplify separation, and the market value of by-products may be increased by conversion.

Beside the main product fibres the pulping industry produces a black liquor containing lignin, hemicellulose, hydroxycarboxylic acids and carboxylic acids with a high water content. After thickening, this effluent is mainly used for thermal energy production via incineration. The volatile carboxylic acids are removed with the water during thickening. Separation of acids from the condensate is recommended for environmental reasons. State of the art recovery of carboxylic acids from aqueous effluents is reactive extraction followed by distillation for solvent regeneration. However, distillative solvent regeneration produces an azeotropic mixture of carboxylic acids and water causing energy intensive product isolation. Therefore, reactive distillation was investigated to intensify solvent regeneration and product isolation. Methanol was added as esterifying agent to the laden solvent containing formic acid and acetic acid, converting the carboxylic acids into their corresponding low boiling methyl esters. Esterification improves the separation of the acids and the reactive extractant and simplifies separation of the distillate products due to lower boiling points of the methyl esters compared to the acids. A homogeneous (4-dodecylbenzenesulfonic acid) and a heterogeneous catalyst (Amberlite IR120) were tested to speed up the reaction rates in reactive distillation. The admixture of 5 w% of 4-dodecylbenzenesulfonic acid and a proton equivalent amount of Amberlite IR120 in batch reactive distillation experiments provided complete conversion of formic acid and 88 % yield of methyl acetate in the distillate. High solvent regeneration efficiency at low temperature and ambient distillation pressure is achieved. The heterogeneous catalyst enables simple separation from the solvent in the reboiler. In homogeneous catalysis catalyst separation step may be challenging. To circumvent an additional separation step the influence of the homogeneous catalyst in the solvent phase on extraction efficiency of acid extraction was investigated. Loss of solvent load capacity is of minor impact. Due to the preference of the reactive extractant to strong acids, the catalyst 4-dodecylbenzenesulfonic acid is kept in the solvent phase during extraction without losses. During distillation the catalyst-extractant aggregate is cleaved at operation temperature in the reboiler and therefore available for catalysis of esterification, enabling combination of extraction and reactive distillation without catalyst separation.

568081 Jet Fuels Production from Waste Cooking Oil Via Catalytic Transfer Hydrogenation

The progressive depletion of fossil fuels and the environmental impacts associated with such exploitation has fostered research on alternative and renewable fuel sources. Currently, the transportation sector, especially the aviation industry, relies on liquid hydrocarbon-based fuels which requires kerosene-range hydrocarbons with strict quality standards to power gas-turbine engines. These hydrocarbons can be obtained from renewable sources, such as triglycerides. Waste cooking oil (WCO) is a suitable renewable fuel and is not an edible source like virgin vegetable oil. It can represent a sustainable alternative feedstock as it is widely produced globally. Moreover, its use in jet fuels production converts waste into a valuable product resulting in a circular economy.

In this work, a novel process for jet fuels production from WCO (catalytic transfer hydrogenation (CTH)) is thoroughly investigated. CTH utilizes hydrogen-donating compounds and a suitable catalyst to release H_2 *in situ* which saturates and reduces triglyceride molecules. CTH is potentially beneficial relative to conventional hydrogenation because it avoids the handling of large quantities of H_2 gas at high operating pressures. Jet fuels production from WCO was investigated experimentally in a continuous-flow lab-scale fixed-bed reactor, using 2-propanol as hydrogen donor and a granular activated carbon catalyst. The experiments were carried out at low pressure (2 bar) and temperature (300-400°C). The reaction products were fully characterized using GC-MS. An optimum yield of 72 wt% liquid hydrocarbon products was obtained at 380°C.

Using the experimental results obtained in the lab, a detailed techno-economic analysis (TEA) and life cycle assessment (LCA) were performed to assess the profitability and environmental impacts of CTH. Both processes were simulated via Aspen Plus® process simulation software to determine material and energy balances. For the LCA, a "well-to-pump" system boundary is used, incorporating all material and energy flows during resource extraction up to the delivered product (i.e. jet fuel). A functional unit of 1,000 ton-d^{-1} WCO feedstock was employed for the life cycle inventory (LCI). Characterization factors were obtained from Argonne's GREET model and SimaPro's Ecoinvent database to determine three (3) key environmental impacts: (1) total energy use (including fossil and petroleum), (2) greenhouse gas (GHG) emissions, and (3) energy return-on-investment (EROI). For the TEA employing a 25-year life, cost information was collected to determine capital cost, direct cost, indirect cost, fixed operating cost, variable operating cost, and energy cost. Modeling results for this holistic system is benchmarked against a conventional hydrogenation process to fully capture the novel process' overall sustainability.

The results of the analysis highlight that although both processes achieved comparable jet fuel/diesel product yields, CTH would require significantly lower capital and energetic costs. However, the cost of the 2-propanol feedstock might hinder the process profitability under current market scenarios. CTH's "well-to-pump" total energy consumption is 25% lower than conventional hydrogenation, for all products and co-products. CTH also reflects a 35% GHG reduction compared to conventional hydrogenation. Both processes were highly-sensitive to the cost of 2-propanol. While the market price for 2-propanol is significantly variable, this study shows that CTH could potentially be a promising, greener and safer process for jet fuels production from WCO. This work demonstrated that with reduced feedstock supply costs, the tradeoff between environmental impacts and profitability of CTH-based jet fuels is achieved.

562866 The Conceptual Framework and Modeling Analysis of Compact Energy Systems

The conceptual frame work for compact energy system is presented to make energy from a biomass like wood/switchgrass, as an alternative to the construction of large scale biorefineries. The compact energy system consists of four individual components: biomass preparation, biomass converter, fuel reformer, and powered-engine. The individual unit processes within each component and the possible types of compact energy systems with different biomass converter technologies like fermentation, pyrolysis, and gasification are presented. The size, weight, and energy efficiency of upgrading biomass to energy using compact energy system with various gasification technologies have estimated. For a compact energy system with the hydrogen fuel cell as a powered-engine component and that processes 10 kg of dry biomass per day, the net energy (kWh) of compact energy system is found to be -7.5, -30, 18.7, 13.1, and 11.7 with the super-critical, microwave assisted, catalytic, steam, and conventional gasification technologies as biomass converter technologies, respectively. The low yields of super-critical gasification and a low efficacy of converting electric energy to heat via electromagnetic waves with the microwave assisted gasification are results in such negative net energy with the respective compact energy systems. Finally, the challenges and opportunities with the development of compact energy systems built around gasification are discussed.

575367 Synthesis and Evaluation of Dipicolinic Acid (DPA) – Based Interwell Tracers for Reservoir Surveillance

Our efforts to address the known deficiencies of modern-day chemical tracing in hot and high salinity reservoirs has led to the adaptation of developed technology such that it touches upon multiple facets of tracer management, with particular emphasis on materials development and their subsequent detection. Fundamentally, this technology comprises classes of optically detectable and chromatographically distinguishable molecular tracers that serve as selective chromophore ligands for trivalent lanthanide metals capable of sensitizing their luminescence upon ligand-metal ion complexation, thus enabling high-sensitivity detection. Lanthanide-based chelation chemistry already has a known basis in applications involving the detection of biological entities; in particular, dipicolinic acid (DPA), a biomarker of bacterial spores, exhibits high-specificity binding to sensors containing Tb^{3+}/Eu^{3+} metal ions. In continuing to exploit the optical properties of such systems, we herein report the synthesis of DPA-derived molecular inter-well water tracers, placing particular emphasis on backbone functionalization of native DPA to generate "barcodes" that can enable their simultaneous full-field deployment. Preliminary coreflood results benchmarking DPA against conventional fluorobenzoic acid (FBA)-based and ionic (KCl) tracers indicate that DPA exhibits similar breakthrough times, is non-retentive and also non-partitioning. We have also demonstrated that DPA and its variants are amenable to chromatographic separation and detection by high performance liquid chromatography (HPLC), providing ultra-trace limits of detection (LOD) using a method complementary to optical detection modalities. Studies regarding the physicochemical properties of the lanthanide ion complexes generated are being conducted concurrently alongside evaluation of synthesized tracer candidates, the integrity of which are also being evaluated via long-term thermal stability testing. We hope that the research conducted herein will ultimately lead to the deployment of new classes of chemical tracers capable of facilitating increased waterflood efficiency, particularly in mature reservoirs where optimization of oil recovery techniques requires high levels of detail.

573670 Design Development of a Novel Sour Water Stripper

With an increasing focus on energy performance improvements and strict environmental laws, there is a need to recycle water resources and reuse them in the industry. Refineries consumes huge amounts of water which offer an opportunity to reprocess it in an efficient manner. This study focusses on the design and simulation of sour water stripping (SWS) unit in the refinery setup for the reuse of sour water. The stripper removes the H_2S and NH_3 content from the sour water making it recyclable for various processes. In this study, process simulation has been performed to analyze the energy saving and cost reducing prospects in the sour water stripping unit. First, a base case design has been simulated using a reference sour water stripping unit data. The developed model is validated against the plant data in order to ensure the model validity. Various alternative designs have been developed to seek process improvements either by improving the energy efficiency or by reducing the sour water stripping unit cost. All the process designs were compared in terms of energy requirement and economics while maintaining the stripped water purity of 10 ppm H_2S and 100 ppm NH_3 content.

A novel design has been proposed in this study that can significantly reduce the energy requirement compared to the conventional designs. The proposed design employs mechanical work using a compressor instead of hot and cold utilities. In this design, the stripped water stream leaving at the bottom of the column is flashed to a reduced pressure, whereby, some amount of vapors are generated. The produced vapors leave at the top of the flash drum and are fed to the two-stage compression train. A compression ratio of three is assumed in each compressor stage with an inter-cooler. The compressed vapor stream at 2.1 bar and 240.7 °C then heats up the incoming feed stream and is fed to the bottom of the column at 121.7 °C.

The results of this study provide a meaningful analysis in terms of energy and economic considerations to improve the performance of the sour water units in the refineries. The results show that the proposed vapor recompression design can reduce the hot and cold utilities requirement by 90 % and 22 % respectively compared to the base case design. Despite the fact that vapor compression design requires higher initial capital investment because of expensive compressors, the economic analysis shows that over the 25 years plant life, the operating energy savings are substantial which makes vapor compression design more economical. The results reveal that the vapor compression design requires 53 % less operational energy compared to the base case design which translates into the annual operating cost savings of more than 60 %. This results in total annual cost of M$ 3.09 for the vapor compression design compared to the M$ 6.49 for the base case design.

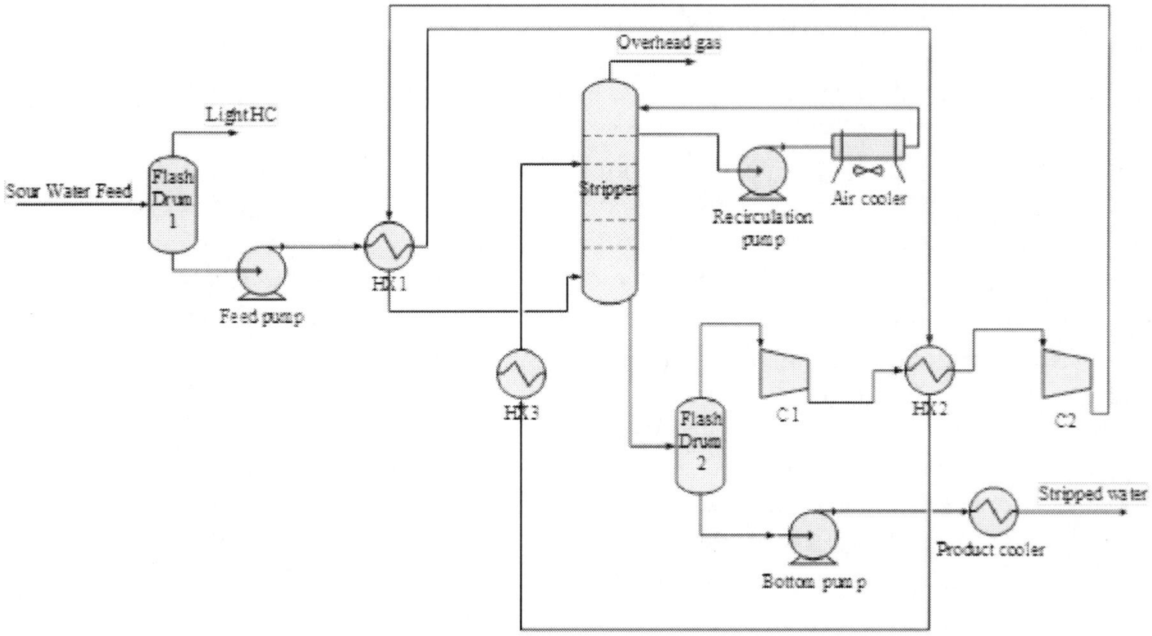

Figure 1: Process flow scheme for the proposed vapor compression design

578760 Photoelectrochemical Sensor

The evolution of photoelectrochemical (PEC) bioanalysis has resulted in substantial progress in its analytical performance and biodetection applications. PEC sensor represents a unique means for chemical and biological detection, with foci of optimizing semiconductor composition and electronic structures, surface functionalization layers, and chemical detection methods. Here, I have briefly discussed my recent developments of nanowire-based PEC sensing, with particular emphasis on three main detection mechanisms and corresponding examples. I have also demonstrated the use of the PEC sensing of real-time molecular reaction kinetic measurements, as well as direct interfacing of living cells and probing of cellular functions. My work will serve as a useful source to inform the interested audience of the latest developments and applications in the field of PEC bioanalysis.

580152 High-Throughput Sensing of Single-Bacterium Growth: Toward Rapid Antibiotic Susceptibility Testing

Accurately measuring the number of viable microorganisms plays an essential role in microbiological studies including antibiotic susceptibility testing (AST). Since conventional culture-based AST methods are too time-consuming (typically 24-72 h), rapid AST is urgently needed for preventing the increasing emergence and spread of antibiotic resistant infections. Although several phenotypic antibiotic resistance sensing modalities are able to reduce the AST time to a few hours or less, concerning the biological heterogeneity, their accuracy or limit of detection are hampered by low throughput. Here, we present a rapid AST method based on whole slide imaging (WSI)-enabled high-throughput sensing of single-bacterium growth, conducted on custom-built on-glass-slide culturing device. The time for determining the minimum inhibitory concentration (MIC) was theoretically shortest that ensures the growth of each individual cell present in a large population is inhibited. Reliable MIC of *E. cloacae* against gentamicin was obtained within 1 h, while the gold standard broth dilution method required at least 16 h for the same result. Owing to the feature of high-throughput quantitative analysis, our approach can be further applied to timely identifying the rising antibiotic resistance adapted to the treatment and guiding the adjustment of treatment strategy to prevent the aggravation of antibiotic resistance.

570379 Ionophore-Based Optical Sensors Incorporating Long-Lifetime Phosphorescent Microparticles for Background-Free Ion Detection in Biological Samples

Fluorescent sensors are ideal for biological applications due to their low invasiveness and miniaturizability. However, signal interference from autofluorescence and scattering limits their practical application. Long-lifetime phosphorescent materials have emission lifetimes on the order of hours or even days, and have long been used commercial in street signs and 'glow-in-the-dark' materials. More recently these materials have attracted interest as agents for biological imaging and sensing as a method for reducing background noise, since, by programming a delay between excitation and emission collection, the short-lived signals from scattering and autofluorescence can be avoided completely. In this work we demonstrate an ionophore-based optical sensor incorporating long-lifetime phosphorescent microparticles for background-free sodium detection in biological samples. We couple the phosphorescent microparticles to a pH-sensitive quencher dye and sodium-binding ionophore to create a dynamic and reversible optode film selective for sodium. We also characterize the sensor in terms of selectivity and stability and demonstrate its usefulness to detect sodium levels in human serum.

576752 Quantification of Biomarkers at Point of Need Using Metal Enhanced Fluorescence and Surface Acoustic Waves

Point of need (PON) quantification of cancer biomarkers such as those used in monitoring disease progression requires sensing at low (pg/ml) limits of detection from body fluids such as blood. For example, proteins such as CEA and CA19-9 are monitored by clinicians for progression of pancreatic cancer, and CA125 is monitored for ovarian cancer progression. Proteins such as PSA are used for initial screening for prostate cancer. All such tests are currently performed in clinical laboratories, off site. The capability to provide such monitoring at the clinician's office or hospital is challenging, as these proteins need quantification at low, pg/ml to ng/ml levels from a body fluid such as blood, which contains interfering proteins at orders of magnitude higher concentrations. We present concepts and results for such detection using plasmonic (metal) enhancement of fluorescence for lowering detection limits, and Rayleigh surface acoustic waves (SAWs) for reducing interference from non-specific binding (NSB) of other proteins in an immunofluorescence format. These same Rayleigh SAWs are utilized in our platform to reduce incubation time, by inducing sample mixing.

In metal enhanced fluorescence (MEF), fluorescence intensity is increased by metallic nanostructures in the proximity of the fluorophore, leading to lower limits of detection. Recent methods for creating metallic nanostructures, such as e-beam nanolithography, colloidal lithography, and colloidal self-assembling, require complicated processes and have various drawbacks. In this work, we have developed a simple nanostructure fabrication process combining thermal annealing and oxide deposition. A thin silver film is treated with rapid thermal annealing to generate the nanostructures on a SAW device surface. These nanostructures are then coated with a thin silica film to protect them and also to control the distance between the metallic structures and fluorophores. This process provides a solution to achieving silver nanostructures over a large area with simple processing techniques. Immunofluorescence experiments conducted using these surfaces show significant enhancement of fluorescence intensity. The effect of initial silver film, and oxide coating thicknesses on fluorescence enhancement was studied in detail. All fabricated nanostructures showed enhancement, with the silica layer producing an additional enhancement. The optimized structure produced a significant intensity enhancement of ~19X, leading to the construction of an immuno-sensor with a limit of detection of a few 10's of pg/ml. Finite difference time domain (FDTD) simulations for the enhancement factor allowed for understanding the role of the silica coating, and optimization of the immuno-sensor.

Acoustic streaming from Rayleigh SAWs is utilized to remove NSB proteins to improve sensor signal in our PON sensor platform. We show that NSB proteins which interfere with sensor signal are removed effectively, and incubation times are reduced significantly utilizing these SAWs. The sensor electronics and optics for this proposed sensor platform can be easily reduced to a small, hand-held package, which we present, with further reduction in size possible. The platform is demonstrated with quantification of a typical biomarker, CEA, from blood samples.

586770 Chemically & Electrochemically Interactive Characterization of Biomaterials *in vitro*

The ever growing demand for advanced and high performance implantable biomedical materials entails development of better interactive in vitro characterization tools to obtain a more representative prediction of their performance in vivo. Herein cases of implantable biomaterials, for temporary bone scaffolding support and coronary stents applications have been discussed.

Temporary implants pose a great challenge where the implant is expected to degrade and dissolve after a period of time at which point it is either absorbed by the body or excreted in the urine. The two specific medical applications are the temporary scaffolding for bone fixation and the coronary stents application. This is to prevent the long term complications associated with presence of a foreign material in the body, for example inflammatory reactions, thrombosis and restenosis in stent applications associated with the permanent stainless steel stents. Research in this area has revolved around magnesium in the past decade as it is safely metabolized by human body upon degradation and yet possesses adequate mechanical and load-bearing properties to support a healing bone as a scaffold or an artery from elastic recoil after balloon angioplasty. The main shortcoming however is the rapid degradation of magnesium in physiological environment that causes premature failure of the implant. In a series of studies [1–7] magnesium alloys were designed and fabricated, the underlying degradation mechanism in simulated biological environment was unraveled and surface treatments were applied to fine-tune degradation rate and biocompatibility. Scanning electrochemical microscopy was extensively used as a uniquely interactive characterization technique to examine degradation of magnesium at high spatial resolution in-situ. A new bio-sensing method was introduced for detection of H_2 which enabled detection of localized degradation. Localized degradation is one of the main reasons for the premature failure of load-bearing structures. The new insight into the degradation mechanism allowed for an informed exploration of surface treatments to fine-tune degradation rate of magnesium in a physiological environment. In particular, ionic liquid surface treatment, conversion coatings based on rare earth metal and a biodegradable conducting polymer coating were investigated in order to suppress the initial rapid degradation, alkalization and H_2 evolution while in the longer term creating a self-healing surface film capable of maintaining uniform degradation pattern. Results of these works introduced candidates with optimized bio-degradation rate that would minimize the risk of early stage alkalization and inflammatory reaction for the surrounding tissue as well as minimizing risk of premature failure to ensure adequate mechanical support is provided during the healing (of the bone) or remodeling (of the artery).

In the field of biomaterials sciences, although there are standards developed to examine the performance of materials in certain in vitro environments, there is still major need for a testing procedure with holistic approach to take into account most environmental parameters, e.g. load and cycles, ions/electrolyte, proteins, cell adhesion, mobility, temperature etc. Currently the in vitro test can be considered only as a preliminary screening test with limited ability to give the realistic picture, and a holistic simulation procedure has to be developed to measure performance in the biological condition. For example a hip and knee simulators with the tribological test properties of the materials together with full consideration of the corrosivity in the cell culture environment and other interacting elements. It is only with a holistic approach and using a multi-array of sensors that the full impact of an implant material on the host can be determined. Surface modifications are often performed on the biomedical implants to improve corrosion resistance, wear resistance, surface texture and biocompatibility. The interactions that take place at the atomic level between the surface of the implant, the host and the biological environment including all types of micromotions of the implants similar to what is expected in the human system should be studied thoroughly in order to advance implant materials with predictable function in the body.

References

[1] S.S. Jamali, S.E. Moulton, D.E. Tallman, M. Forsyth, J. Weber, G.G. Wallace, Applications of scanning electrochemical microscopy (SECM) for local characterization of AZ31 surface during corrosion in a buffered media, Corros. Sci. 86 (2014) 93–100.

[2] S.S. Jamali, S.E. Moulton, D.E. Tallman, M. Forsyth, J. Weber, G.G. Wallace, Evaluating the corrosion behaviour of Magnesium alloy in simulated biological fluid by using SECM to detect hydrogen evolution, Electrochim. Acta. 152 (2015) 294–301.
[3] S.S. Jamali, S.E. Moulton, D.E. Tallman, M. Forsyth, J. Weber, A.S. Mirabedini, et al., Corrosion protection afforded by praseodymium conversion film on Mg alloy AZNd in simulated biological fluid studied by scanning electrochemical microscopy, J. Electroanal. Chem. 739 (2015) 211–217.
[4] Y. Zhang, X. Liu, S.S. Jamali, B.R.W. Hinton, S.E. Moulton, G.G. Wallace, et al., The effect of treatment time on the ionic liquid surface film formation: Promising surface coating for Mg alloy AZ31, Surf. Coatings Technol. 296 (2016) 192–202.
[5] S.S. Jamali, S.E. Moulton, D.E. Tallman, Y. Zhao, J. Weber, G.G. Wallace, Self-healing characteristic of praseodymium conversion coating on AZNd Mg alloy studied by scanning electrochemical microscopy, Electrochem. Commun. 76 (2017) 6–9.
[6] D. Liu, G. Xu, S.S. Jamali, Y. Zhao, M. Chen, Fabrication of biodegradable HA/Mg-Zn-Ca composites and the impact of heterogeneous microstructure on mechanical properties, in vitro degradation and cytocompatibility, Bioelectrochemistry. 129 (2019) 106–115.
[7] S.S. Jamali, S.E. Moulton, Y. Zhao, S. Gambhir, M. Forsyth, G. Gordon, A biodegradable conducting polymer coating to mitigate early stage degradation of magnesium in simulated biological fluid : an electrochemical mechanistic study, ChemElectroChem. 6 (2019) 4893–4901.

578352 Solvent-Resistant Electrochemical Transistors As Dual Sensing Platforms in Aqueous and Organic Media

Organic electrochemical transistors (OECTs) are promising contemporary sensing platforms for high-fidelity transduction of (bio)chemical signals into electrical observables. In an OECT, such signals cause can ions to inject from an electrolyte into an organic semiconducting polymer channel, changing its bulk conductivity and the output drain current. Such devices provide multiple advantages that make them apt for bioelectronics applications, namely (1) operating voltages less than 1 V, (2) direct interfacing with biologically relevant electrolytes for *in vivo* application, and (3) printable fabrication, for flexible and stretchable form factors. Despite such advantages, OECTs have only achieved high signal transduction with a narrow subset of channel materials, many of which require cytotoxic cross-linkers to prevent dissolution in aqueous environments and/or exhibit poor processability in carrier solvents. Consequently, there is a strong need to develop biocompatible channel materials with high-processability.

In this work, I discuss the development of single-component OECTs which are insoluble to both aqueous and organic media, yet can be facilely solution-processed from commercial water-soluble carboxylated polythiophene (CPTs) precursors post-synthesis. Such OECTs demonstrate robust stability and high pseudocapacitances even after repeated 2-V width voltage sweeps, as well as high transduction (transconductances = 30 mS at V_{gate} = -0.5 V) in aqueous and nonaqueous systems. Changes in doping state and ion injection will also be probed using spectroelectrochemistry and electrochemical quartz crystal microbalance to further understand the origin of high transconductance in these materials. In conjunction with other electrochemical techniques, we should be able to elucidate detailed characteristics that dictate high mixed ionic-electronic transport in these materials.

563640 Investigation of Polymer-Plasticizer Blends for BTEX Sensing Using Acoustic Wave Devices

Volatile organic compounds (VOCs) are chemicals which evaporate easily and become gases at ambient temperature and pressure. The EPA has made the monitoring of VOCs mandatory due to their health impacts, which range from headaches and nausea to cancer. The current techniques used for VOC monitoring such as photo ionization, gas chromatography, mass spectroscopy, and e-noses sensors are expensive, time-intensive, demand rigorous sample preparation and cannot quantify the VOCs in the air. To employ the shortcomings of current sensor systems, quartz crystal microbalance (QCM) and surface acoustic wave (SAW) sensors based on polymer-plasticizer thin films have been developed. Acoustic wave devices are capable of detecting very low concentrations (ppm, ppb) of organic compounds in the environment. Imprinted polymers as sensing film have been used to detect BTEX gas analytes to a lower limit of 0.5ppm using SAW devices.

Traditionally, the sensitivity and performance of rubbery polymers towards hydrocarbon sensing are superior to glassy polymers however, they show low molecular selectivity due to their amorphous nature, availability of large free volume elements, and display of undesirable viscoelastic effects. Plasticization helps to modify the chemical and physical properties of the glassy polymer by decreasing the glass transition temperature, altering the pore dimensions and increasing the free volume of the polymer film thus enabling higher diffusion and sorption of the analyte molecules. This work shows that the selectivity and sensitivity of glassy polymers can be tailored by introducing a plasticizer. Poly (ethyl methacrylate) (PEMA), Poly (methyl methacrylate) (PMMA), PEMA-PMMA, PMMA-Polystyrene copolymers were modified by introducing plasticizers, diisononyl cyclohexane-1, 2-dicarboxylate (DINCH), diisooctyl azelate (DIOA), dibutyl sebacate (DBS), n-butyl stearate (BS), and di-n-butyl phthalate (DBP) to enhance BTEX sensitivity and selectivity in presence of interferants.

The objective of this study is to investigate the effect of various plasticizers on glassy polymers sensitivity by unraveling the polymer-plasticizer transient response mechanism and identifying the accurate plasticizer for superior BTEX sensing. This work demonstrates that the incorporation of suitable plasticizers into a polymer-based sensor can vastly improve the selectivity for BTEX compounds based on properties such as sorption (via thermodynamic partition coefficients), saturation dynamics, stability, and viscoelasticity of the film. A mathematical model will be incorporated to select the optimum polymer-plasticizer blend for BTEX sensing in gas and liquid phase using QCM and SAW devices.

575123 Microfabricated Silion Nanowire Array Platform for Surface Enhanced Raman Spectroscopy

This work presents a microfabricated silicon nanowire array chip for Surface Enhanced Raman Spectroscopy (SERS) to rapidly detect trace target analytes in biofluids such as serum, urine and exhaled breath. Silicon nanowire array are created by wet etching and coated with silver nanoparticle. Silver nanoparticle coated silicon nanowire array chips are characterized for surface enhanced Raman spectroscopy (SERS). The silicon nanowire array can result in significant SERS for target analyst. The chips are used to detect trace THC in exhaled breath by SERS. The results show that silicon nanowire array coated with sputtered silver nanoparticles can achieve a limit of detection of 3.1 pg of THC. The linear relationship between SERS signal and the amount of THC indicate that the method is suitable for quantification of THC concentration in breath.

579139 Findings from MTO Commercialization – Don't Forget about the Small Stuff

In the rush to commercialize a new technology, there are many things to consider: reactor scale up & design; engineering design of major equipment; simulations; thermodynamics; sales tools; customer discussions, and a lot more. One of the things that can be easily missed is what happens to components that are produced in minor concentrations. It may be easy to ignore these when a simulation indicates that they will be washed away in waste streams (or maybe become product impurities). However, these minor concentration components can become problematic when they don't respond exactly as predicted, especially when they build up their concentrations in recycle loops. This talk will discuss some these type of issues that have occurred with MTO commercialization and what we are doing about them.

573116 De-Risking Technology Development and Scale-up Efforts Effectively in Fuel and Chemical Processes

The sequential pathway for de-risking new technologies includes bench scale testing, pilot programs, demonstration plants and finally commercial plant construction. In the past, hastening, or eliminating any of the above steps would increase the risk to novel technologies, and to the companies that invest in them. Unfortunately, this pathway requires significant time and financial investment. In the past few years, in response to increased pressure to speed up the pathway, engineers have developed many tools and strategies for mitigating that risk. Modern process simulation, targeted piloting programs, toll processors, directional technology development plans and modular strategies can all help companies achieve commercial success from technology development through scale-up to commercialization. When employed properly, these, and other tools can decrease project risk, shorten the time to commercialization, and limit the overall cost of developing innovative technologies. This presentation will discuss how to utilize these, and other tools to quickly and effectively de-risk process technology development. Using case studies of past projects, it will also review the process of developing technologies from bench scale through pilot and demonstration scale to full commercialization within the framework of these de-risking tools. Case studies include next-generation biodiesel development, conversion of natural gas liquids into valuable chemicals and marine fuel, and production of high purity alumina from waste streams.

565316 Can Outsourcing R&D Balance Myopic Business Management?

Abstract

When Ford Motor Co.'s CEO, Jim Hacket, announced plans in April 2018 to reduce spending by $25.5 billion over the next four years [1], Ford staff were concerned about how the pending cuts would impact R&D investments [2]. Similarly, DowDuPont Inc.'s CEO, Ed Breen, announced the elimination of R&D spending on "moonshots" to focus on projects that cost less than $30 million to complete and, perhaps most importantly, carry a lower risk profile [3]. From one perspective, these announcements are merely part of normal operations, as business leaders make decisions to increase, reduce, or cut spending every day. On the other hand, these announcements raise a few questions:

- Were these part of a trend in R&D spending?
- What are some of the factors that affect CEO decision-making?
- What are some consequences of myopic decision-making?
- Can outsourcing R&D balance myopic business management?

After briefly addressing these questions, this presentation will disclose a relevant case study in which MATRIC, a contract R&D organization, successfully helped a client to develop and commercialize a new technology.

References

[1] A.Levine-Weinberg, "Ford Reveals More Details on its Massive Cost-Cutting Plan," Fox Business, August 13, 2018 (https://www.foxbusiness.com/markets/ford-reveals-more-details-on-its-massive-cost-cutting-plan).

[2] L.Brooke, "Underway on Nuclear Power," Automotive Engineering, June 2018, p. 18 (http://www.nxtbook.com/nxtbooks/sae/18AUTP06/index.php#/22).

[3] J.Kaskey, "DowDuPont CEO Kills Off R&D 'Moonshots' in Bid for Higher Profit," Bloomberg, May 31, 2018 (https://www.bloomberg.com/news/articles/2018-05-31/dowdupont-s-new-profit-formula-killing-more-research-projects).

555687 Succeeding at Innovation: Generating Value from New Technology at Each Stage of Development and Implementation

Succeeding at Innovation: Generating Value from New Technology at Each Stage of Development and Innovation

Technology provides increased capability, energizes our imagination, and leads to business advantages. Too often our embrace of technology is weakened by narrow and inaccurate measures of value, the challenge of changing existing practices, and the inertia of our own minds and customs.

Examining the effect new technology has on our systems at each stage of its development and implementation reveals the full value of its effect. The potential gains at each stage, how to map an uncertain path to the future, and methods to decide the level and timing of investment are discussed in the context of drug product delivery, processing, and scale up. Examples from modeling and simulation, 3-D printing, and novel lyophilization technology are used as illustrations.

552804 Heat Transfer Coefficient in Two Phase Liquid-Solid Fluidized Bed

Heat Transfer Coefficient in Two Phase Liquid-Solid Fluidized bed

Marwa Sabah Mahdi ,Dr.Jamal M.Ali* and Dr.Ameer A.Abdul-Rahman

University of Technology Department of chemical Engineering, Iraq

***80002@uotechnology.edu.iq**

Abstract

Hydrodynamic of liquid,-solid fluidized bed was studied experimentally as well as theoretically.The impact of the superficial velocities of liquid, particle diameter ,the liquid holdup, solid holdup, height of bed and minimum velocity fluidizing were investigated. Individual phase holdups were determined through static pressure approach from the information of the pressure drop, the bed height and the properties of the solid and liquid.

Heat transfer coefficients to a liquid-solid fluidized bed in a cylindrical tube have been measured using water as liquid phase and three sizes of spherical glass beads particles as solid phase and electrically heated single vertical tube heated electrically with different heat transfer rate (500, 600 and 700W). The independent variables included heat flux, liquid velocity and particle physical properties. The results showed that the heat transfer coefficient increased with increasing fluidizing water velocity and also with increasing heat flux. It was also determined that the heat transfer coefficient increased with increasing the diameter of solid particles and decreased with increasing the voidage fraction. On the other-hand, the heat transfer coefficient changed with the change of positions of heater distance from the distributer. The experimental results as well as a data bank containing a large number of measured heat transfer coefficients for solid- liquid fluidization over a wide range of operational parameter have been compared with the predictions of most published correlations. The results were agreed in compare with the existing correlations predictions. These correlations were found to be approximate in reproducing the data of present work.

Key word: liquid-solid, fluidized bed, minimum fluidization velocity- solid hold up, heat transfer

571285 Design and Optimization of 3D Reactor Technologies for the Production of Light Olefins

Roughly 1.5×10^8 t of ethylene and 8×10^7 t of propylene are produced annually worldwide. Although some alternative technologies have emerged and/or gained interest over the last decade, the bulk of these molecules are and will continue to be produced through the steam cracking process. [1]

In this process, a hydrocarbon feedstock is heated together with steam in a tubular reactor to temperatures of up to 900 °C. To reach these high temperatures, the reactors are typically suspended in a gas-fired furnace. Due to these elevated temperatures and large scale of the process, it is one of the most energy-intensive processes in chemical industry, accounting for an annual energy consumption of about 3.0×10^{18} J, [2] which is the equivalent of approximately 500 million barrels of oil or twice the annual electricity consumption of Germany. Next to this, the process is also responsible for the emission of nearly 200 million tons of CO_2.

As the process has already been established for a long time, a lot of effort has been put into increasing the process efficiency, leading to a high level of complexity and limited room for further process improvements with conventional optimization techniques. In this respect, Djokic et al. [3] propose technological innovations to push beyond contemporary conventional steam crackers. These innovations range from alternative fuels such as oxy-fuel over material improvements such as high-emissivity coatings or advanced high temperature alloys to the enhancement of convective heat transfer from the reactor coil to the process gas by implementing 3D reactor technology.

These 3D reactor technologies not only reduce the fuel consumption and lead to longer coil lifetime by reducing the tube metal temperature (TMT), they also decrease coke formation which increases the process run length. This is mainly achieved by two methods: on one hand, the reactor wall shape can be altered to increase the heat exchanging surface. On the other hand, inserts or wall modifications can be introduced to increase the degree of mixing. This can be achieved at a large scale by radial mixing, i.e. secondary motions displacing hot fluids away from the reactor wall into the colder center of the flow, and on a smaller scale by increasing the turbulence in the near-wall region. This near-wall turbulence increase, will decrease the thermal boundary layer, which is the region in which the largest resistance to heat transfer can be located.

Even though a large number of heat transfer enhancing designs have been proposed and/or implemented in the field of heat exchangers, the number of designs suited for high temperature applications such as steam cracking is rather limited. Several reasons can be mentioned for the limited range of designs; specialized coil materials do not easily allow complex deformations or 3D patterns; turbulators can increase the risk of coke depositions clogging the reactor and the use of these designs inevitably leads to an increase in pressure drop over the reactor, which might negatively affect the yield of the desired light olefins.

Comparisons between different designs in an industrial setting are not widely available. To cope with this, flow, heat transfer and chemical kinetics are now modeled by means of Computational Fluid Dynamics, providing a unique insight into the effect of reactor design on flow behavior and heat transfer. This modelling effort is validated based on the results of an extensive experimental campaign, where different enhancement techniques are investigated with stereo-PIV, for flow field measurements, as well as heat transfer measurements, based on liquid crystal thermography (LCT). These experiments provide detailed local information and hence allow a thorough comparison with the simulated flow field and heat transfer capabilities of these designs.

The modelling methodology is used to investigate the performance of two recently developed reactor technologies namely SFT[4, 5] and SCOPE[6]. These technologies are then compared against three established reactor designs; finned[7] and ribbed[8] coils as well as coils with IHT inserts[9]. Simulation results will focus on pressure drop, tube metal temperatures, product distribution and coking rates, allowing a unique comparison of the most relevant industrial reactor designs.

[1] I. Amghizar, L. A. Vandewalle, K. M. Van Geem, and G. B. Marin, "New Trends in Olefin Production," *Engineering,* vol. 3, pp. 171-178, Apr 2017.

[2] T. Ren, M. K. Patel, and K. Blok, "Steam cracking and methane to olefins: Energy use, CO2 emissions and production costs," *Energy,* vol. 33, pp. 817-833, May 2008.

[3] M. R. Djokic, K. M. Van Geem, G. J. Heynderickx, S. Dekeukeleire, S. Vangaever, F. Baffin-Leclerc, *et al.*, "IMPROOF: Integrated Model Guided Process Optimization of Steam Cracking Furnaces," *Sustainable Design and Manufacturing 2017,* vol. 68, pp. 589-600, 2017.

[4] C. M. Schietekat, M. W. M. van Goethem, K. M. Van Geem, and G. B. Marin, "Swirl flow tube reactor technology: An experimental and computational fluid dynamics study," *Chemical Engineering Journal,* vol. 238, pp. 56-65, Feb 2014.

[5] M. W. M. van Goethem and E. Jelsma, "Numerical and experimental study of enhanced heat transfer and pressure drop for high temperature applications," *Chemical Engineering Research & Design,* vol. 92, pp. 663-671, Apr 2014.

[6] Schmidt+Clemens. (2018, 31/01/2018). *Strengthen the heart of your plant! SCOPE Fusion HTE - the steam cracker tubing of the future*. Available: https://www.schmidt-clemens.com/products/scope/

[7] C. M. Schietekat, D. J. Van Cauwenberge, K. M. Van Geem, and G. B. Marin, "Computational Fluid Dynamics-Based Design of Finned Steam Cracking Reactors," *AIChE Journal,* vol. 60, pp. 794-808, Feb 2014.

[8] D. J. Van Cauwenberge, J. N. Dedeyne, K. M. Van Geem, G. B. Marin, and J. Floré, "Numerical and experimental evaluation of heat transfer in helically corrugated tubes," *AIChE Journal,* 2017.

[9] A. Carrillo, F. Bertola, G. Wang, and L. Zhang, "Intensified heat transfer technology–CFD analysis to explain how and why IHT increases runlength in commercial furnaces," in *AIChE and EPC 2010 Spring National Meeting, San Antonio, TX*, 2010.

572006 Control of Reactive Distillation Columns Separating Reacting Mixtures with Somewhat Unfavorable Relative Volatilities

For the separation of reacting mixtures with somewhat unfavorable relative volatilities (i.e., the reactants are the light and heaviest components and the products are the lightest and heavy ones), arranging an external recycle from the bottom to a certain stage of the reactive section onto the reactive distillation columns (RDC-ERBR) can highly improve the system steady state performance. In this article, the dynamics and control of these novel process designs separating a hypothetical reversible reaction, A + B ↔ C + D ($a_C > a_A > a_D > a_B$), were studied. With the arrangement of the external recycle, the severe asymmetry between the two product compositions and the most lightest reactant feed flow rate can be suppressed as compared with conventional reactive distillation columns, thereby presenting favorable influences to process dynamics and controllability. Three decentralized control systems with different manipulating modes of the external recycle flow rate (i.e., it is set to be a constant value, ratioed to B feed flow rate, and used to manipulate the B composition in the sensitive stage) were developed for RDC-ERBR to evaluate its influences to the process control. The results showed that the decentralized control system of the RDC-ERBR with the external recycle flow rate ratioed to B feed flow rate has the best closed loop control performance as compared to the other decentralized control systems of the RDC-ERBR and the CRDC.

567645 Intensification of Cephalexin Synthesis in Modular Microfluidic Devices Using Electric Field

Cephalexin is an important ß-lactam antibiotic that is enzymatically synthetized from a nucleophile (7-aminodeacetoxycephalosporanic acid – ADCA) and an acyl donor (phenylglycine methyl ester – PGME). Cephalexin is thermodynamically unstable and is typically produced in the kinetic regime. Prolonged exposition of cephalexin to the enzyme results in cephalexin decomposition to ADCA. Hence, good control of reaction conditions is necessary to obtain high cephalexin yield. The problems accompanying the cephalexin synthesis can be partially avoided by using microfluidic platforms.

Cephalexin synthesis can be carried out in a homogeneous arrangement with a soluble enzyme with high catalytic activity. Cephalexin extraction from the reaction mixture limits the successive hydrolysis and allows for the enzyme reuse. Aqueous two-phase systems (ATPS) represents an enzyme-friendly alternative to the extraction provided by organic solvents. In this work, we used ATPS consisting of polyethylene glycol (PEG), phosphate salts and water. We have identified such ATPS composition that is characterized by high affinity of the top PEG phase to cephalexin and the reactants and by very low affinity to the enzyme.

In the next step, we have optimized properties of the reaction mixture to attain high cephalexin concentration in the course of the enzyme reaction. We focus on the effects of temperature, the concentrations of ADCA and PGME and their molar ratio. We identified regimes that lead to ADCA conversion to cephalexin higher than 50 % within a few tens of minutes in a batch arrangement. The batch experiments gave us the estimate on the residence time of the reaction mixture in flow-through microfluidic arrangements.

Two microfluidic reactors-separators for the cephalexin synthesis have been developed. One of them relies on the slug flow arrangement of immiscible aqueous phases. The bottom phase contains the enzyme. The reactants are introduced in droplets of the top phase. The reactants are transferred through the interface into the bottom phase where cephalexin is synthetized. Cephalexin is simultaneously transferred back to the top phase where it is stabilized due to the absence of the enzyme. After phase separation, the bottom phase with the enzyme is led back to the start of the process.

The other arrangement, which is currently tested in our lab, contains a long microcapillary that serves as flow-through reactor. Both reactants and the enzyme are introduced in one phase. The output is led to a parallel flow microextractor where cephalexin and unreacted ADCA and PGME are transported to other phase. Enzyme in the original phase is led back to the microcapillary inlet. Moreover, the parallel flow microextractor is equipped by two electrode chambers. We have found that low intensity electric field applied perpendicularly to the interface allows for 100% separation of the product from the enzyme phase. Parametrical studies relevant to both arrangement will be presented.

Selected References:

Cech, J.; Hessel, V.; Pribyl, M., *Chem Eng Sci* **2017,** *169*, 97-105.
Cech, J.; Pribyl, M.; Snita, D., *Biomicrofluidics* **2013,** *7* (5), Artn 054103.
Cech, J.; Schrott, W.; Slouka, Z.; Pribyl, M.; Broz, M.; Kuncova, G.; Snita, D., *Biochem Eng J* **2012,** *67*, 194-202.
Vobecka, L.; Romanov, A.; Slouka, Z.; Hasal, P.; Pribyl, M., *New Biotechnol* **2018,** *47*, 73-79.
Pribyl, M., *Sep Purif Technol* **2019,** *221*, 311-318.

Acknowledgement: The authors thank for the support by the grant of the Czech Science Foundation [grant no. 17-09914S].

571470 Immobilisation of Eosin Y on Wool for the Photocatalytic Synthesis of Fine Organic Chemicals

Recently, the interest in solar photocatalysis for the synthesis of organic compounds has grown as a more sustainable alternative than traditional synthesis routes. Moreover, transition metals have been replaced by metal-free dyes, which are cheaper and non-toxic. In this field, Eosin Y has shown a great potential as a photocatalyst for applications in organic synthesis reactions such as cyclisation, dehydrogenation or redox reactions, among others.[1] However, these reactions have been performed with the catalyst in solution, which increases the cost of the process due to the difficulty to handle and separate the catalyst from the reaction mixture.[2] The immobilisation of Eosin Y on a robust/renewable material allows the continuous illumination of the catalyst during the reaction and its recovery and reuse, making the process more sustainable.

Natural wool was chosen as support of Eosin Y and different immobilisation methods were investigated. The oxidative homocoupling of benzylamine was selected as a model reaction and it was performed under a solar light simulator (40 cm*40 cm). The results demonstrated that acidic conditions are the most favourable to immobilise Eosin Y. This is due to the negative charge of Eosin Y and the wool in water. Treating the wool with an acid creates a positively charged wool surface and an ionic bonding between the negative charge of the catalyst and the wool surface. The stirring of the reaction mixture and low volumes achieved a conversion of 39 % after 5 h. A reusability study was performed, showing that Eosin Y was stable on the wool after 3 cycles. Further investigations have been carried out pre-treating the wool with a cationic reagent to increase the amount of catalyst immobilised, achieving conversions of 43 %. The pre-treatment allowed the increase of positively charged functionalities on the surface of the wool and therefore more ionic bonding sites were created for the catalyst. Further work will be to scale up the reaction with the Eosin Y on wool using a Spinning Mesh Disc Reactor. In summary, this research has demonstrated that the immobilisation of Eosin Y on wool under acidic conditions creates a strong bonding between the catalyst and the support and it is a cheap and sustainable catalyst to carry out photocatalytic synthesis of fine organic compounds.

REFERENCES

1. Prasad Hari and B. König, *Chem. Commun,* 2014, **50**, 6688-6699.
2. Shivaprasad, M. D. Jones, D. A. Patterson and E. A. C. Emanuelsson, *Chem. Eng. Process.*, 2017, **122**, 550–559.

572170 Towards Sustainable Industries: Industrial Symbiosis of an Oil Refinery and a Petrochemical Plant

A common factor that can be seen in the history of all manufacturing and production sectors is that they all go through significant changes to keep them operating efficiently, thus a constant cycle of system development, obsolescence and advancement is formed. These sectors must be able to maintain the balance between the energy consumption and efficiency to keep the system optimised and preserve the opportunity to create value. Oil refineries are a prime example of one of these sectors that play an important role in our day-to-day lives. Due to an oil refineries' major dependency on crude oil, the price fluctuation of which, has a profound impact on this industry. Thusly, this drives up prices and forcing a rise in the search for alternative solutions to tackle these efficacy and economic problems. Therefore, in this work, an integration approach has been considered based on the Industrial Symbiosis concept to connect an oil refinery with a petrochemical plant.

One main feature of the larger chemical industry that should be considered, is that they require careful management of material and energy to obtain valuable products. Of course, the processes key to the generation of the valuable products also produce fewer desirable chemicals, which have no value as waste products. Some of these waste materials can be used as feed-stock for other processes; turning probable costly chemical disposal into an economic boon. This "greener" approach opens avenues of improvement compatible with the idea of industrial symbiosis; where the waste from one process can be a feed-stock, useful for another. Taking this into account, a case study is presented involving the connection of an oil refinery with an ethylene production plant through material exchange and stream combination to benefit both plants.

As a result, we were able to improve not only product quality but the overall profit of both plants by a significant margin, while also decreasing dependency on outside sources for material supply.

575632 Selection of Isolation Methods for Micron Sized Particles

Solid-liquid separations are commonly used in industrial processes and are approached in a distinctive way based on the desired component- solid or liquid. The separation technique is dependent on a combination of factors- how the feed suspension was prepared, total volume, solid concentration, and particle shape and size distribution of the solids. Investigating the performance of various separation equipment is time consuming and this search can be narrowed down by utilizing specific laboratory scale tests to predict the behavior of a separation technique before scale-up. Herein, the processes of technology selection for situations where solids are the desired fraction with particle size < 10 mm are discussed. The choice of cake filtration methods such as vacuum and pressure filtration is often based on the average flux, which indicates the time required for filtration. Filter media with small channel sizes are required to isolate the micron sized particles. The cake deposited always exhibits high resistance and the small particles may block the channels leading to unacceptably long filtering time. Sedimenting centrifuges such as tubular, or stacked disc can be used for isolating sub-micron particles since they do not use a filter medium for separation and avoid the challenges of media fouling. The selection of sedimenting centrifuge can be based on laboratory scale feasibility tests performed with a benchtop centrifuge by varying the centrifugal force and residence time and observing the sediment compactness and supernatant clarity. Other filtration methods including thin cake filtration and cross-flow filtration can also be considered for systems with small particles.

576220 Techno-Economic Assessment of Hydrogen Production from Parallel Design Configuration of Fossil Fuel Derived Syngas Generation Processes

With an increase in the industrialization and population in the last few decades, not only the gap between energy demand and supply has increased but also the CO_2 emissions hit the highest levels. To control the greenhouse gas emissions, the improvement in the power generation processes to produce clean energy with CO_2 control technologies are getting more and more attention. Integrated coal gasification and combined cycle (IGCC) processes is a pre-combustion carbon capture technology that represents higher process efficiencies with the large scale implementation of carbon capture and sequestration (CCS) systems. In this study, coal based gasification systems utilizing the low/high quality coals have been coupled with the reforming technologies in the parallel design configuration followed by the sequential heat integration between the two processes. This integration not only provides the necessary heat from the gasification unit to the reforming unit but also increase the H_2/CO ratio of the syngas. The syngas can be directly burned in the combined cycle unit or converted into the H_2 gas which itself can be marketed as a clean fuel. The techno-economic analysis have been also performed to analyze the energy and cost required for producing syngas and H_2 from the standalone IGCC and reforming technologies and the results are compared with the newly developed design for the various syngas compositions. The results showed that the new design has potential to boost up the syngas/H_2 production capacities and also reduces the CO_2 emissions for unit amount of H_2 production. Furthermore, the co-gasification of coal and biomass in the new design also showed the potential to reduce the overall carbon foot print.

576947 Process Intensification of Biogas to Liquid (BGTL) Processes: A Techno-Economic Study

Process Intensification of Biogas to Liquid (BGTL) processes: A techno-economic study

Benard Nyawanga, Ahmed Naqi (co-author), Babu Joseph, and John Kuhn*

Department of Chemical and Biomedical Engineering, University of South Florida.

**Presenting author. Email: nyawanga@mail.usf.edu , Tel : (954)632 0931*

ABSTRACT

Industrial conversion of biogas to liquid fuel has been a subject of study for some time now. In current technology, biogas is converted into liquid fuel through two major steps: tri-reforming step to produce syngas (a mixture of CO and H2), and Fischer-Tropsch Synthesis (FTS) step to convert the syngas to a spectrum of hydrocarbons [1]. Separation and upgrading of the produced hydrocarbon mixture allows production of synthetic transportation fuels. Currently efforts are underway to create catalysts that can convert biogas directly to liquids in a single reactor. The objective of this paper is to compare the process economics of such an intensified process with the current two step BGTL process.

A detailed modeling of the two-step process is carried out using ASPEN Plus process design software package. Data for the process is based on literature, combined with laboratory results on the biogas to liquid conversion process. The current commercial diesel prices is used to evaluate the potential revenue from selling the product in the open market. The total capital investment to construct the plant with a capacity of handling 100,000 ton per year of wet biomass is $16.2 million with a potential of producing 2.60 million gallons of diesel [2].Corn stover is used as the biomass, which is then converted into biogas through anaerobic digesters. The annual operating cost to run the plant is estimated to be $8.81 million [3]. An annual revenue from selling the diesel product is estimated to be $14.6 million taking into account a green energy incentive of $3.00/gallon of diesel sold. The net present worth at the end of the plant life is $8.76 million with a discounted cash flow of return of 26.2%. The breakeven cost of diesel is determined to be $4.34/gallon assuming no tipping fees are charged for handling the waste [3].

The major challenge of the Industrial two steps method is the cost of production. In this process, there is considerable energy demand from the reformer. FTS produces energy but it is not enough possible to recycle this energy due to the low operating temperature of the FTS unit. The reformer operates at higher temperatures (~1073 K) and lower pressure (~3 bars) while the FTS reactor operates at lower temperature (~700 K) and higher pressure (~10 bars). In the proposed intensified process, both the reformer and the FTS are combined into a single reactor. This way, an average optimal temperature and pressure conditions can be used in the presence of a novel catalyst. The combined process will be modeled using ASPEN Plus software and economic analysis will be done in order to estimate the feasibility of a such a process. A comparison between this new process and the two step process will be done to determine the potential payback from an intensified process.

- Williams. *Bioenergy Conversion Technologies*. Madison, WI: University of Wisconsin. Accessed Date:02/18/2018.Available:http://www.wgbn.wisc.edu/conversion/bioenergy-conversion-technologies
- EIA. (2018). *Weekly Retail Gasoline and Diesel Prices*. Accessed Date: 02/10/2018. Available: https://www.eia.gov/dnav/pet/pet_pri_gnd_dcus_nus_m.htm.
- A. Naqi, "Conversion of Biomass to Liquid Hydrocarbon Fuels via Anaerobic Digestion: A Feasibility Study" Masters of Science in Chemical Engineering M.S. Thesis, Department of Chemical and Biomedical Engineering, University of South Florida, Tampa, Florida, 2018.

577746 Life Cycle Analysis of the Implementation of CO_2 Capture Process in Power Plants: Effect of Type of Fuel and Energy Demand

Because of current high concentrations of greenhouse gases (GHGs) in the atmosphere, global warming is one of the main challenges to solve in the 21st century. The burning of fuels generates around 32 Gton of carbon dioxide per year, 40% are related to the generation of electricity and about 20% to the operation of the chemical and petrochemical industry. Although CO2 cannot be eliminated as a product of combustion, it is possible to reduce CO2 emissions with the implementation of different technologies named CO2 capture or transport and storage (CCS). Among the different technologies for the CCS, post- combustion CO2 capture using alkanolamines as a solvent, is the most developed technology and is also considered to be the most feasible method. In recent years, several authors have emphasized the need to carry out a comprehensive assessment of the capture process, since a wider vision prove if the energy used during the regeneration of solvents, as well as in the transport and storage of CO2, the environmental benefit can be minimal or even zero. Some studies have been reported the Life Cycle Analysis (LCA) as a quite useful tool to evaluate the environmental impact of different CO2 capture systems. In general, it has been shown that in order to reduce the environmental impact through the use of these capture systems, initially it is essential to improve the efficiency of the process during the generation of electricity since the coupling of the capture plant can significantly reduce the efficiency of the combined cycle. In the present work, the environmental impact of coupling a CO2 capture plant to an electric power generation plant was evaluated. The use of four different fuels in the power plant was considered; biogas, coal, non- associated natural gas, and an associated natural gas. Two operating scenarios were considered; in the first, the same fuel flow was considered for all the plants and in the second, the same energy demand was specified. The design and simulation of the process plants were developed through the use of the ASPEN Plus simulator, while the LCA was carried out with the SIMAPro software.

563391 Systematic Design of Carbon Constrained Industrial Parks

Natural gas, the cleanest fossil fuel, is an vital resource for the oil and gas industry as it is used both as raw material and an energy source. It is considered the transition fuel to a carbon free economy and is key material for many economies. Different routes exist to monetize natural gas to value added products and each route comes with corresponding carbon footprint and cost. Effective natural gas monetization is of increasing importance especially with the global consensus on emission reduction and visions towards carbon free economies. It poses challenges the industry to adhere to emission reduction targets, while maintaining profitability. Most emissions due natural gas stem for stationary industrial emissions such as stacks, gas turbine exhausts and byproducts (waste) from processing. The stationary nature of the emissions however, gives an advantage to capture large amounts of emissions within close proximity as oppose to scattered mobile emission e.g. transportation emissions. Thus, monetization decisions of natural gas with carbon reduction were investigated. The method that explores natural gas usage options either as conversion to value added products such as liquefied natural gas (LNG), Gas-to-Liquid (GTL), Methanol, etc or use as fuel in power plant under emission targets. The industrial city was taken as a system and modeled through balances and constraints, which were optimized applying deterministic solvers. The systematic, optimization-based approach to simultaneously determined natural gas monetization and carbon dioxide management through carbon capture, utilization and storage (CCUS) as well as renewable energy strategies. It integrates energy in both heat and power between different processing plants within the industrial city boundary using a central utility system. The holistic method compared the different natural gas value added products routes from economics (capital and operating costs) with its carbon dioxide emissions that were either eliminated through the use of renewable energy or by converting carbon dioxide in the CCUS network. The approach could determine the investment needed, the capacity, allocation and pipeline network design of the materials exchanged. This framework enables the assessment of different climate reduction policies, which is a powerful tool for regulators and industrial cities designers. Several case studies were solved that considered an existing industrial city that needed to meet emission requirement and the design of a new grass-root city with limited flow of natural gas and an imposed emission target.

571355 CFD Study of Water Spray Cooling System for Refinery Process

This study focuses on the optimization of water spray cooling system used for air the cooling system of a commercial refinery company by both experiments and computational fluid dynamics (CFD). The cooling performance of air cooling system used to decrease in summer because of increase in the atmospheric temperature. But the air cooling system with water spray could effectively maintain its cooling performance using high sensible and latent heat of water. Although water spray cooling system has higher performance than air cooling system, the system could be corroded if the water mist dose not evaporate. It is necessary to optimize the water spray conditions so that the water used is little enough to be evaporated.

The CFD simulation study of air cooling system using horizontal type water spray has been analyzed through various studies. The vertical type spray, which is also used in this study, has originality of research owing to the rarity. Main parameters of study were spray angle, rotation speed of fan, water quantity, spray type, and droplet size. Finally, an optimized water spray cooling system that has the best cooling performance without corrosion could be designed.

Modelling of Anaerobic Digestion of Wastes From Industrial Processes

Pooja Sharma*, Uttam Kumar Ghosh, Amiya Kumar Ray*****

***Assistant Professor, Desh Bhagat University, Mandi Gobindgarh, Punjab, India**
****Associate Professor, Indian Institute of Technology, Roorkee, India**
*****Retired Professor, Indian Institute of Technology, Roorkee, India**

ABSTRACT

Generation of wastewater and sludges from various municipal and industrial activities require stabilization of organic pollutants prior to disposal into the environment or for further reuse. Biological treatment processes, anaerobic digestion, in particular mostly adopted for treatment of high organic strength wastes such as municipal sludges and industrial wastewater. The process performance in terms of percentage of Chemical Oxygen Demand (COD) removal of wastes depends on the degradation of organic pollutants by microbial species and their survival conditions. Several substances present in the wastes which inhibit microbial growth kinetics or instabilize process conditions resulted in digester failure. Some of the factors which instabilize process conditions of anaerobic digester are considered in the present investigation. The microbial growth studies have also been carried out under steady state or transient conditions by microbial growth kinetic models. The inhibitory effects on performance of digester by the presence inhibitory substances in terms of high influent substrate concentration are also considered. Simulation of the developed model under transient conditions can quantify the extent of process stability which is considered to be of paramount importance in selecting a suitability of a process for specific applications.

Keywords: Anaerobic Digestion, Process Model, Stability Enhancement, Growth Kinetics

INTRODUCTION

Effluent treatment plant is an imperative necessity for cleaning water bodies from industrial discharges to protect the environment from pollutants. The major industries include chemical process industries, refineries, pulp and paper, sugar, distilleries, tannery, olive pomace, mining, metallurgical, brewery and many others. The wastewater generated from these industries is highly organic in nature. Anaerobic treatment technology generally known as biological treatment is usually used to treat solid wastes or excess sludge from aerobic processes and also wastewater generated from various municipal and industrial activities. Out of the numerous equipments the most popular and economically feasible, technology is anaerobic digestion, mostly preferred by industry.

Modelling is very important in understanding the interrelation of different parameters pertaining to the system under consideration (sewage sludge and distillery waste) and therefore optimization followed by control. For anaerobic treatment various kinds of models are available. These are kinetic models, transport phenomena models based on equation of continuity and population balance models to quantify the population of microbial species.
There are many models available in literature for different types of waste in anaerobic digester. These are due to Hill *et al.* (1977); Havlik *et al.* (1984); Moletta *et al.* (1986); Smith *et al.* (1988); Kiley *et al.* (1996) and many others. The studies on model stability enhancement by various researchers (Topiwala *et al.*, 1971; Durate *et al.*, 1981; Bello-Mendoza *et al.*, 1998; Azeiteiro *et al.*, 2001; Beneventi *et al.*, 2009). Every developed model requires validation from the experimental data or data available from literature. Research on the development of mathematical models for anaerobic digestion process is going on by considering various aspects. The present investigation has been planned to develop dynamic models.

MODELLING AND SIMULATION

A large body of literature is available on kinetic and dynamic models describing anaerobic digestion process. Due to the complexity of the biochemical reactions involved, this process most routinely suffers instability problems. Majority of the researchers in this field considered methanogenesis as the rate limiting step. Methanogens are highly sensitive to parameters such as pH, temperature and overload of organic material under transient conditions and presence of inhibitory compounds, which results in instabilized process conditions. In order to enhance stability, a better understanding of process operation and development of proper controllers for optimization of the process, various steady state or dynamic models have been developed over the years using various microbial growth kinetics. The developed kinetic models may be classified into three broad categories (Lyberatos *et al.*, 1998; Gavala *et al.*, 2003) such as:

- Models describing inhibition by unionized volatile fatty acids
- Models describing inhibition by total volatile fatty acids

- Model describing inhibition by NH_3 as key parameter

The first model describing the inhibitory effect of substrate on growth kinetics of microorganisms was due to Haldane model Eq. 2.1 (Bailey *et al.*, 1987; Jana *et al.*, 2008).

$$\mu^{M} = \frac{\mu_{M}^{M}}{1 + \frac{K_{S}^{M}}{S_{1}} + \frac{S_{1}}{k_{i}}} \quad (2.1)$$

where,

μ_{M}^{M} Maximum specific growth rate (g/g)

μ^{M} Specific growth rate (g/g/day)

S_1 Concentration of inhibitory substrate (g/L)

k_i Inhibition constant (g/L)

K_S^A Half-saturation constant (g/L)

The model due to Haldane explains well the inhibitory effect of substrate at higher concentration and growth limiting characteristics at lower concentration.

The dynamic model development for biological processes started in 1970s. At that time due to inadequate knowledge of process and unavailability of modern computer operation, the solution of systems of nonlinear differential equations of the dynamic models was difficult. Andrews (1969) developed first dynamic model for anaerobic digestion process was based on Haldane model, considering the inhibitory effect of unionized volatile fatty acids (e.g., acetic acid) which acts as precursor for methane formation on methanogens. Further model due to Andrews *et al.* (1971) considered only methanogenesis step for dynamic model development as expressed in Eq. 2.2.

$$\mu^{M} = \frac{\mu_{M}^{M}}{1 + \frac{K_{S}^{M}}{HVA_{1}} + \frac{HVA_{1}}{k_{i}}} \quad (2.2)$$

where,

HVA_1 Unionized volatile fatty acid concentration (g/L)

The above dynamic model considered three phases, liquid, gas and biological. The model was validated by performing laboratory studies and also based on previously published data. Attempts have been made to develop various models with Andrews (1969) model as basis; these are due to Hill *et al.* (1977); Havlik *et al.* (1984); Moletta *et al.* (1986); Smith *et al.* (1988); Kiley *et al.* (1996) and many others. Many other researchers considered two microbial groups viz., acidogens and methanogens (without restricting to methanogens which was advocated by Andrews *et al.*, 1971) for anaerobic digestion modelling based on different kind of raw materials. Further modification has been made by Smith *et al.* (1988) by introducing hydrolysis step in formulating more general dynamic model. All the above models (except Haldane) were common regarding the inclusion the inhibitory effect of substrate and methanogenesis step as rate limiting step.

Various researchers (Hill *et al.*, 1977; Moletta *et al.*, 1986; Havlik *et al.*, 1986; Kiley *et al.*, 1996) considered ammonia inhibition step in the model development of anaerobic digestion process for complex organic substrates. Kiley *et al.* (1996) developed dynamic model considering ammonia inhibition during anaerobic co-digestion of municipal solid waste and sewage sludge. Inhibitory effect of ammonia on methanogens was observed resulting in process failure. All the models (Hill *et al.*, 1977; Havlik *et al.*, 1986; Moletta *et al.*, 1986) were however as a modified version of Andrews (1969) model. All the aforesaid models were validated through laboratory experimental data. Angelidaki *et al.* (1993) proposed a model, relatively complex in nature, which also considered ammonia inhibition during degradation of insoluble and soluble carbohydrates.

Some other factors have also been considered by other researchers (Bryers *et al.*, 1985; Gavala *et al.*, 2003) during development of respective models using different composition of wastewater. Several modifications have been put forward over the years by several investigators such as Sotemann *et al.* (2005); Batstone *et al.* (2006); Lubken *et al.* (2007); Hafnera *et al.* (2009); Uysal *et al.* (2010); Rapport *et al.* (2011); Appels *et al.* (2011); Yu *et al.* (2013). It is important to point out that the dynamic and kinetic models proposed by all the investigators mentioned above were based on different kind of raw materials and different factors pertaining to the problem.

Experimental studies were carried out and mathematical models were developed considering ammonia inhibition on anaerobic digestion using different kind of wastes, notably animal waste, sludges, cattle manure, poultry waste,

piggery waste etc. There are plenty of literature available on ammonia inhibition on laboratory studies as well as model development.
Angelidaki *et al.* (1994) proved through experiments that during anaerobic digestion of cattle manure the failure of process occurs at concentration of ammonia at 700 g/L. Gallert *et al.* (1997) reported that 50% reduction in process performance at a concentration of ammonia 560-568 mg/L during anaerobic digestion of piggery waste. Sung *et al.* (2003) carried out experiments on synthetic wastewater. In their investigation digester was operated at five different concentration of total ammoniacal nitrogen ranging between 0.4 g/L to 5.77 g/L. Performance of digester with respect to COD removal and methane production rate at different concentration were predicted. At the total ammoniacal nitrogen concentration of 5 g/L corresponded to 12.4 L/day of methane flow rate whereas percentage COD removal was found to be 61.2%. When the digester was operated at concentration of 6 g/L, reduction in percentage COD removal from 61.2% to 46.8% was achieved. Methane flow rate decreased from 12.4 L/day to 7.32 L/day. Sung *et al.* (2003); Cali *et al.* (2005) described the failure of anaerobic digester due to toxic effect of ammoniacal nitrogen. The inhibitory range of ammoniacal nitrogen causing digester failure was lied in between 1700 mg/L to 14000 mg/L as mentioned by Espana-Gamboa *et al.* (2012).
Chen *et al.* (2008); Henjfelt *et al.* (2009) addressed the issue regarding the ammonia inhibition and concluded that this is dependent upon the nature of substrate to be digested anaerobically and period of acclimatization. Experimental results presented by Ansari *et al.* (2012) on distillery spent wash (based on cane molasses) were reported with influent having ammoniacal nitrogen concentration of 1254 mg/L and after anaerobic treatment the concentration in effluent was found to be 714 mg/L. Literature studies also shows the inhibition cause by ammoniacal nitrogen in the distillery waste obtained from raw materials such as grain, maize, cereals, etc.
According to Hobson *et al.* (1976), digester should be operated at higher concentration of NH_3-N if the bacterial species is properly acclimatized. Successful operation has been observed at ammoniacal nitrogen concentration of 2700 mg/L found by step increase in NH_3-N values instead of gradual increases and slight inhibitory effect on performance was observed at NH_3-N concentration of 3000 mg/L during anaerobic digestion of piggery waste.
Yenigun *et al.* (2013) discussed the importance of ammoniacal nitrogen as nutrient for microorganisms, which act as buffer to resist digester stability at highly acidic conditions. An excellent review made by Rajagopal *et al.* (2013) described the importance of ammonia in enhancing digester stability as a buffer at a low concentration, also acting as nutrient, which is beneficial for microbial growth rates. Higher concentration caused inhibitory effect on microorganisms. They further suggested that in order to operate digester at high ammoniacal nitrogen concentration, the inoculum in the digester should be highly acclimatized. The author has classified the effect of varying concentration of ammoniacal nitrogen on digester performance and suggested the important criteria as under: beneficial effect if ammoniacal nitrogen concentration lies in between 50-200 mg/L; no antagonistic effect in the range between 200-1000 mg/L; inhibitory effect predominates between 1500-3000 mg/L and induces strong inhibitory effect if it is above 3000 mg/L.

A comprehensive study has been made on the various models pertaining to inhibition caused by some key factors such as unionized volatile fatty acid, total volatile fatty acids and ammonia on digester performance and stability at high loading rate conditions. A brief report on available literature related to anaerobic digestion of industrial effluent (distillery) and sludge (sewage sludge), effect of wall growth factor and available kinetic model for anaerobic digestion process, finally, influence of microbial kinetic parameters on process performance are cited. Further the experimental studies carried out by various researchers on waste degradation were also reported. Regarding enhancement stability of digester under transient conditions and at high loading rates done by previous researchers were however restricted to sewage sludge. However, effect of wall growth in gaseous phase was not considered in previous studies.

DISCUSSION

Development of dynamic models for anaerobic digestion process of distillery waste with particular reference of those due to Andrews et al. (1971) and Hill et al. (1977). The shortcomings of previous models are indicated. Based on the suggested methodologies put forward by the above investigators attempt has been made in this present study to develop modified dynamic model using some relevant assumptions permitted for industrial exploitation. Influence of wall growth factor has been discussed for continuous operation using transient conditions. Simulation of the model has been carried out to examine the effect on various parameters, such as , , alkalinity, etc. Inhibitory effect of unionized volatile fatty and ammonia on methanogens growth kinetics was considered at higher influent substrate concentration. Model validation with industrial data has been undertaken along with sensitivity analysis. Digester instability conditions have been pointed out in terms of percentage COD removal and methane production.

CONCLUSIONS

The present investigation pertains to anaerobic digestion of wastes of industrial importance. These studies involve development of steady state kinetic model, as well as dynamic models for the above two types of wastes. Preliminary data for industtrial waste was collected from industry from both inlet and outlet of anaerobic digester followed by their laboratory analyses. These have been used for simulation purpose. Published data are used for model validation.

REFERENCES

1. Acharya, B. K., Mohana, S., and Madamwar, D. (2008). Anaerobic treatment of distillery spent wash - A study on upflow anaerobic fixed film bioreactor. Bioresource Technology, 99: 4621-4626.
2. Andrews, J. F., and Graef, S. P. (1971). Dynamic modelling and simulation of the anaerobic digestion process. Advances in Chemistry, 105: 126-162.
3. Arnaiz, C., Buffiere, P., Lebrato, J., and Moletta, R. (2007). The effect of transient changes in organic load on the performance of an anaerobic inverse turbulent bed reactor. Chemical Engineering and Processing: Process Intensification, 46: 1349-1356.
4. Chen, Y., Cheng, J. J., and Creamer, K. S. (2008). Inhibition of anaerobic digestion process: a review, Bioresource Technology, 99: 4044-4064.
5. Clement, S., Gouiller, A., Ottenio, P., Nivelon, S., Huber, P., and Nortier, P. (2011). Speciation and super saturation model in papermaking streams. Process Safety and Environmental Protection, 89: 67–73.
6. Costello, D. J., Greenfield, P. F., Lee, P. L. (1991). Dynamic modelling of a single-stage high-rate anaerobic reactor - I. Model derivation. Water Research, 25: 847-858.
7. Dyka, C. T., and Saigal, S. (2005). On the development of a hybrid particle method using the differential formulation. International Journal for Computational Methods in Engineering Science and Mechanics, 6: 21-29.
8. B. K. (2003). Kinetics and modeling of anaerobic digestion process. Advances in Biochemical Engineering/Biotechnology, 81: 57-93.
9. Hill, D. T., and Barth, C. L. (1977). A dynamic model for simulation of animal waste digestion. Water Polltion Control Federation, 49: 2129-2143.
10. Hobson, P. N., and Shaw, B. G. (1976). Inhibition of methane production by *Methanobacterium fomicicum*. Water Research, 10: 849-853.
11. Jimenez, A., Borja, R., Martin, A., and Raposo, F. (2006). Kinetic analysis of the anaerobic digestion of untreated vinasses and vinasses previously treated with *Penicillium decumbens*. Journal of Environmental Management, 80: 303-310.
12. Larsen, D. H., and Dimmick, R. L. (1964). Attachment and growth of bacteria on surfaces of continuous-culture vessels. Journal of Bacteriology, 88: 1380-1387.
13. Levenspiel, O. (2004). Chemical reaction engineering John Wiley& Sons, Asia Private Limited, Singapore.
14. Lopez, I., Passeggi, M., and Borzacconi, L. (2015). Variable kinetic approach to modeling an industrial waste anaerobic digester. Biochemical Engineering Journal, 96: 7-13.
15. Lubken, M., Wichern, M., Schlattmann, M., Gronauer, A., and Horn, H. (2007). Modelling the energy balance of an anaerobic digester fed with cattle manure and renewable energy crops, Water Research 41: 4085-4096.
16. Lyberatos, G., Skiadas, I. V. (1999). Modelling of anaerobic digestion - A review. Global Nest: The International Journal, 1: 63-76.
17. Munson, R. J., and Bridges, B. A. (1964). Take over - An unusual selection process in steady-state cultures of *Escherichia coli*. Journal of General Microbiology, 37: 411-418.
18. Neyman, J. (1937). Outline of a Theory of Statistical Estimation Based on the Classical Theory of Probability, Philosophical Transactions of the Royal Society A 236: p.333-380.
19. Normak, A., Suurpere, J., Orupold, K., Jogi, E, and Kokin, E. (2012). Simulation of anaerobic digestion of cattle manure. Agronomy Research Biosystem Engineering, 1: 167-174.
20. Nwabanne, J. T., Onukwuli, O. D., and Ifeakandu, C.M. (2004). Biokinetics of anaerobic digestion of municipal waste. International Journal of Environmental Research, 3: 511-516.
21. Prakash, N. B., Sockan, V., and Raju, V. S. (2014). Anaerobic digestion of distillery spent wash. ARPN Journal of Science and Technology, 4: 134-140.
22. Pullammanapallil, P., Owens, J. M., Svoronos, S. A., Lyberatos, G., and Chynoweth, D. P. (1991). Dynamic model for conventionally mixed anaerobic digestion reactors. AIChE Annual Meeting, USA, 277(c): 43-53.
23. Rajagopal, R., Masse, D. I., and Singh, G. (2013). A critical review on inhibition of anaerobic digestion process by excess ammonia. Bioresource Technology, 143: 632-641.

24. Rapport, J. L., Zhang, R., Jenkins, B. M., Hartsough, B. R., and Tomich, T. P. (2011). Modelling the performance of the anaerobic phased solids digester system for biogas energy production, Biomass and Bioenergy, 35: 1263-1272.
25. Ram, C., Sharma, C., and Singh, A. K. (2014). Electrochemical corrosion investigations on anaerobic treated distillery effluent. Journal of Materials Engineering and Performance, 23: 3321-3327.
26. Seth, R., Goyal, S. K., and Handa, B. K. (1955). Fixed film bio-methanation of distillery spent wash using low cost porous media. Resource, Conservation and Recycling, 14: 79-89.
27. Topiwala, H. H., and Hamer, G. (1971). Effect of wall growth in steady-state continuous cultures. Biotechnology and Bioengineering, 13: 919-922.
28. ZoBell, C. E. (1943). The effect of solid surfaces upon bacterial activity. Journal of Bacteriology, 46: 39-56.

574182 3D Printed Static Mixers for Continuous Reactive Crystallization

Crystallization is an important unit operation in the pharmaceutical industry for separation and purification of intermediate chemical compounds and active pharmaceutical ingredients. The driving force for crystallization is the supersaturation of the solution, which is defined by the difference between the solute concentration and its solubility. There are various methods to create supersaturation. In anti-solvent or reactive crystallization, different process streams are mixed together to generate supersaturation. Consequently, mixing plays an important role in the performance of anti-solvent or reactive crystallization. Initially, mixing is required to obtain a homogeneous solution to obtain predictable bulk nucleation and to prevent local nucleation, which may occur due to concentration gradients[1]. Once crystals are formed, mixing is needed to maintain the crystals in suspension. Plug-flow behavior is typically desired to provide a uniform residence time for each crystal and hence deliver uniform product quality attributes such as crystal shape and size. Tubular crystallizers can provide plug-flow behavior if mixing is fast in radial direction but slow in axial direction. One of the methods to approach such mixing patterns involves static mixers. A kenics-type static mixer has previously been demonstrated to be useful for achieving plug-flow conditions for anti-solvent crystallization of ketoconazole.[2] However, other types of static mixers such has Komax, SMX, or low pressure drop (LPD) mixers have not been studied for their application as a tubular crystallizer.[3] Additionally, the mixing elements of these static mixers can be optimized for a given application.[4] Finally, the mixing requirements may change when the process material flows from the entrance to the outlet of the tubular crystallizer. In particular, in the first part of the crystallizer, rapid mixing of liquid process streams needs to be achieved, whereas large crystals need to be kept in suspension further downstream. Fouling and plugging should be avoided when the crystal mass increases. Therefore, when using a static mixer as a tubular crystallizer, the structure and dimensions of the internals may need to change for optimal operation. These requirements for customization and the absence of any moving parts call for 3D printing to be used as the preferred fabrication method of tubular crystallizers based on static mixers.

The objective of this work is to design, fabricate and characterize tubular crystallizers with different designs for static mixing for reactive crystallization. Stereolithography is identified as the most suitable 3D printing technique among the four major techniques, because it allows for fabrication of leak-proof parts and easy removal of uncured resin from the flow channel.[5] Reactive crystallization of barium sulphate is conducted in a Y-mixer and komax-type mixer fabricated using a commonly available desktop stereolithography 3D printer (Form 2, FormLabs) with clear resin. The results demonstrate practical feasibility of using 3D printing to fabricate innovative tubular crystallizers and a reduced agglomeration is observed in the komax mixer compared to a conventional Y-mixer. Subsequently, reactive crystallization of an active pharmaceutical ingredient, salicylic acid from sodium salicylate and an acid, will be investigated. COMSOL simulations show that a kenics-type static mixer with a standard design provides faster mixing compared to a Y-mixer, komax or a LPD mixer. COMSOL simulations will be used to optimize the different static mixers for the rate of mixing and to minimize zones with low liquid velocity where crystals might settle. Finally, the crystallization performance in terms of yield, crystal size distribution and degree of agglomeration will be characterized experimentally in the 3D printed tubular crystallizer. This work demonstrates a novel type of application of 3D printing in which there is a clear benefit of customization and fabrication speed.

Acknowledgement: The work described in this abstract was supported by a grant from the Research Grants Council of the Hong Kong Special Administrative Region, China, Project No. 16214418.

References

(1) Hacherl, J. M.; Paul, E. L.; Buettner, H. M. Investigation of Impinging-Jet Crystallization with a Calcium Oxalate Model System. *AIChE J.* **2003**, *49* (9), 2352–2362.

(2) Alvarez, A. J.; Myerson, A. S. Continuous Plug Flow Crystallization of Pharmaceutical Compounds. *Cryst. Growth Des.* **2010**, *10* (5), 2219–2228.

(3) Ghanem, A.; Lemenand, T.; Della Valle, D.; Peerhossaini, H. Static Mixers: Mechanisms, Applications, and Characterization Methods - A Review. *Chem. Eng. Res. Des.* **2014**, *92* (2), 205–228.

(4) Singh, M. K.; Anderson, P. D.; Meijer, H. E. H. Understanding and Optimizing the SMX Static Mixer. *Macromol. Rapid Commun.* **2009**, *30* (4–5), 362–376.

(5) Bhattacharjee, N.; Urrios, A.; Kang, S.; Folch, A. The Upcoming 3D-Printing Revolution in Microfluidics. *Lab Chip* **2016**, *16* (10), 1720–1742.

576831 Ultra-Low-Cost Continuous Manufacturing Platforms through 3D Printing and Telescoped Isolation Free Purification

Chemically resistant parts for continuous flow chemistry have been produced using the 3D printing process fused deposition modelling (FDM), from poly(etheretherketone) (PEEK). PEEK has greater chemical resistance than common FDM materials such as acrylonitrile butadiene styrene (ABS), polypropylene (PP), or even high-performance plastics like poly(etherimide) (PEI), and also has good thermal resistance and excellent mechanical strength. These printed parts have been demonstrated to be capable of withstanding high pressures, allowing the superheating of solvents for optimal reaction conditions. The ability to use FDM for these reactors allows highly customisable, cost effective flow reactors to be produced directly in the laboratory and integrated into equipment already available.

The integration of complicated internal structures, difficult to produce with powder bed fusion or resin photo-curing techniques such as selective laser melting (SLM) or stereo lithography (SL), was also possible with FDM, and allowed structures as small as 0.5 mm to be produced reliably within the reactor internal channels. These integrated elements were shown to enhance reaction mixing and the efficiency of liquid-liquid extractions, improving reaction throughput. The printed reactors were used for flow chemistry and then integrated with continuous liquid-liquid extraction followed by continuous phase separation or isolation free continuous crystallisation within an agitated bed crystalliser.

The phase separator is a compact and easily scalable setup, at a fraction of the cost of current commercial systems, requiring only simple back pressure control. The agitated bed allows continuous purification *via* crystallisation on a suspended seed bed, with the option to perform solvent swap as a final step if required. When combined with the printed reactors a customisable synthesis path results, allowing low cost integrated isolation free purification with continuous flow chemistry.

Scheme 1. (Top) 3DP reactor followed by 3DP workup stage and continuous flow phase separation in phase separator. (Bottom) 3DP reactor followed by confined suspension crystalliser

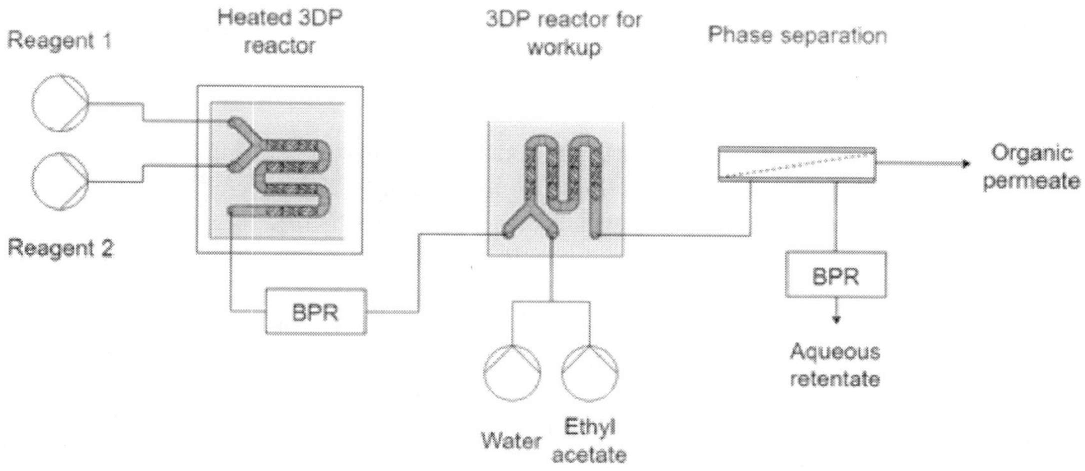
Reagent 1
Heated 3DP reactor
3DP reactor for workup
Phase separation
Organic permeate
Reagent 2
BPR
BPR
Aqueous retentate
Water
Ethyl acetate

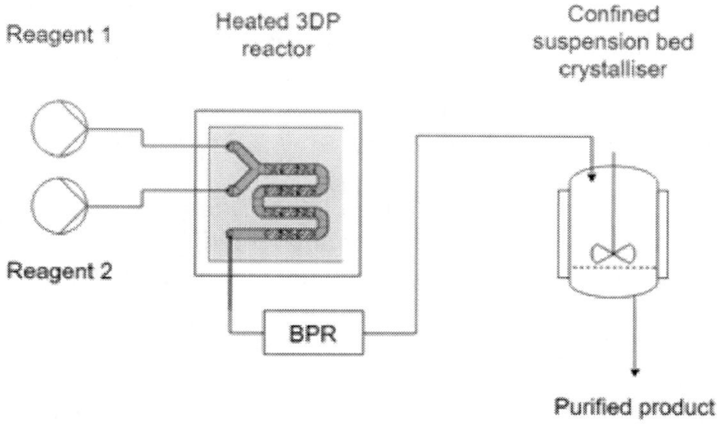
Reagent 1
Heated 3DP reactor
Confined suspension bed crystalliser
Reagent 2
BPR
Purified product

560518 Modeling of Poly(ether imide) in Material Extrusion Additive Manufacturing

Additive manufacturing (AM), is a manufacturing modality that allows the production of small scale, customized parts quickly and cheaply. Material extrusion (MatEx) AM is one of the most popular types of AM. This process uses a filament feedstock to produce a part by depositing thin roads of material in 2D layers which are then stacked in the desired geometry to create a 3D shape. However, this method of manufacturing is currently limited primarily to the rapid production of prototypes and physical models. The ultimate goal of MatEx is to produce parts for the end-use market that will experience practical use. The current state of the process prevents this due to a lack of robust mechanical properties and material versatility. The lack of versatility stems from a highly limited material catalog. MatEx is primarily commercially limited to polylactic acid (PLA) and acrylonitrile butadiene styrene (ABS). This small catalog of materials also limits the mechanical properties of the parts produced with this technique, as the mechanical properties of these two materials are much less than the common engineering thermoplastics such as polyetherimide (PEI) and polyether ether ketone (PEEK). The weakness of the produced parts also stems from a lack of z-axis, interlayer, adhesion. This problem is inherent to the process due to a low amount of time above the glass transition temperature of the material during printing, preventing the diffusion of polymer chains across the interface formed where two roads meet.

To address these issues, we propose a more thorough understanding of MatEx is required in order to more efficiently identify novel materials as well as their optimal printing conditions. Because in-situ measurements of the process are difficult, often resulting in unforeseen changes in the system that render any information gained with the measurements incorrect, we suggest utilizing multiphysics based process models to examine the effect of various process parameters on the quality of the final product. In this study, we developed multiple predictive models based on thermal transport, diffusion, and stress development. By examining the effect caused by modifying various parameters in these models, we can gain a detailed understanding of the process and how it can be modified to allow the printing of novel materials as well as identifying the optimal printing parameters for the production of parts with near bulk properties. These predictive models were then compared to empirical measurements.

Harnessing a High Energy, Superconducting Electron Beam for Additive and Far-from-Equilibrium Manufacturing

Adam Duzik, Ryan Reeves, Justin J. Hill†
Mainstream Engineering Corporation, Rockledge, FL, USA
†Corresponding author: jhill@mainstream-engr.com

Superconducting linear accelerators that generate electron beams open up new opportunities for advanced materials processing since they can continuously deliver extremely high energy electrons. The increased energy delivery rate of these systems enables the production of far-from-equilibrium materials and selective processing within a material surface and subsurface. High energy electrons penetrate deeper within materials and can increase the processing depth and rate. With the support of the Office of Naval Research, NASA, and DoE, Mainstream Engineering commissioned a superconducting linear electron accelerator system specifically designed to investigate advanced materials processing, production, and additive manufacturing. The goal of Mainstream's Electron Beam Enabled Advanced Manufacturing (EBEAM) center is to develop materials that are uniquely produced with high energy electron beam systems and alleviate significant technological deficiencies. We will present results of irradiating various metals and their alloys and the unique material properties that result. Monte Carlo simulations of the interaction of the electron beam with the sample were used to optimize the processing parameters and study far-from-equilibrium processing conditions. Among the applications that will be discussed are additive manufacturing of stainless steels for small modular nuclear reactors and corrosion-resistant cladding via bonding of dissimilar metals for next-generation molten salt reactors.

Introduction

Current manufacturing is subtractive in nature, either removing undesired material from a block of material or joining of sheet metal or simpler components into a more complex whole. These methods result in large amounts of waste and leaving weak weld joints components. Additive manufacturing is an up and coming alternative to subtractive manufacturing. Material is selectively added in 2D layers, resulting in a 3D part. Polymer-based 3D printing as a subset of additive manufacturing is now well-established. Metals and ceramics, however, melt at much higher temperatures, requiring an alternate approach. Metal AM is maturing, with established methods in material feeding and selectively bonding material only in the desired location. Wire-fed systems inject material out of a nozzle, much like 3D printing, while powder bed systems involve spreading successive thin layers of fine powder to construct a 3D build volume.[1] In either case, a metal AM system requires a high intensity, localized heat source to fuse the wire or powder in the desired spots in the final part. Lasers and electron beams are the primary heating methods. Lasers can operate in air, but lack penetration depth, slowing processing speed, and increasing stresses and microfractures.[2,3] Electron beams require vacuum, but can penetrate deeper, melt thicker layers, and can achieve higher power than laser systems. Small parts can be produced in short times, but multiple centimeter-scale parts take several hours, too long for industrial use.

Faster AM times require thicker melt layers without compromising part quality. This in turn requires much faster heating and cooling rates to melt and solidify the powder to solid material in less time. Flash melting to much higher temperatures in turn causes flash cooling and solidification when the beam moves. Rapid cooling creates a strong driving force for nucleation, producing a nanocrystalline structure. As such, higher power localized heating results in faster processing times. To reach more industrially useful times, much higher than current 50-100 kV laser or electron beam systems. Electron beam systems are more amenable to high energy as electrons can reach any energy given enough acceleration. Superconductive resonance cavity linear accelerators are capable of MeV level electron energies, 2 orders of magnitude higher than conventional electron beam AM systems. To this end, Mainstream Engineering Corp. has acquired the Electron Beam Enabled Advanced Manufacturing (EBEAM) system. The superconducting cavity is designed to accelerate an electron beam up to 2 MeV at 4 kW average power, resulting in 10^6 K/s heating and 10^4 K/s cooling rates. Nanocrystalline grain microstructures are accessible at fast raster speeds, enabling larger and stronger parts produced in less time with less waste than conventional manufacturing. Such microstructures are stronger, allowing smaller parts with less material to supports the same stresses as larger parts made with traditional manufacturing. Thus, miniaturization efforts will benefit from such technology.

EBEAM Overview

The EBEAM electron linear accelerator system is shown in Figure 1. The EBEAM system consists of a 60 keV Schottky electron gun and a superconducting linear electron accelerator and currently is capable of producing an electron beam up to 1.4 MeV at 1 kW. Past the accelerator cavity, the electron beam is directed down into a processing chamber; inside, the electron beam strikes component powder material placed onto a build plate. The powder coalesces into a melt pool under electron beam impact. As the beam rasters over the build plate area, the melt pool moves, melting more powder material while the previous material rapidly cools and recrystallizes into solid material.

Additive manufacturing part quality depends on many parameters including: melt pool depth, heating rate, and cooling rate and their roles in controlling solidification and crystallization. Electron beams possess an important advantage over lasers in depth of penetration; this gives electron beams the ability to re-melt and recrystallize larger volumes of already solid parts and materials. Figure 2 shows the recrystallization of existing coarse-grained material into ultra-fine grain material, producing material property advantages, including mechanical yield strength, surface hardness, radiation tolerance and magnetic permeability.[4-13] Such a high powered electron beam also enables new claddings consisting of two dissimilar metals bonded together, enabling high strength and corrosion-resistance even on irregular parts. Such advances are promising for modular nuclear reactor components.

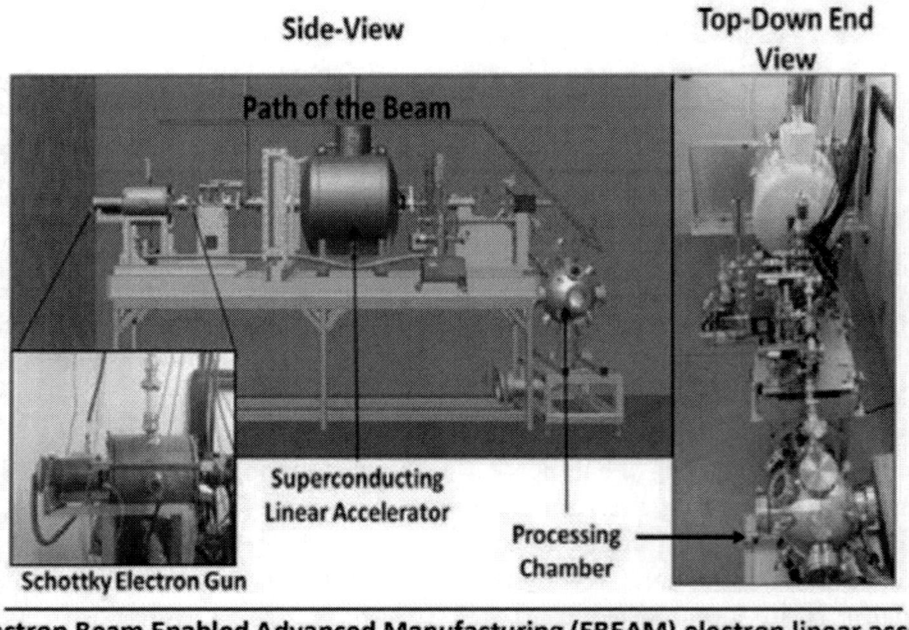

Figure 1. Electron Beam Enabled Advanced Manufacturing (EBEAM) electron linear accelerator.

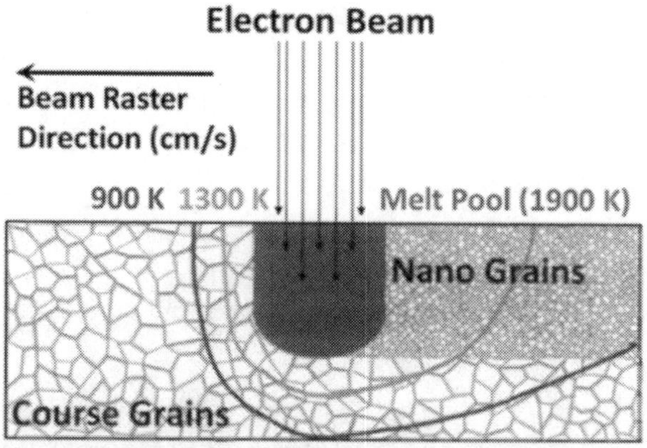

Figure 2. Nanocrystallization process under the higher electron energy possible in the EBEAM system.

Stainless Steel 316L Manufacturing for Small Modular Nuclear Reactors

As developing nations construct a nuclear power infrastructure and aging continues in developed nations' nuclear infrastructure, new generations of nuclear reactors have been proposed. One such concept is the small modular reactor (SMR) that requires far less space than conventional nuclear plants. As a modular design, it is possible to scale these reactors up to meet higher demand and locate the reactors within the consumer site. They can also eliminate power transmission losses and large capital investments needed to build new conventional nuclear plants. Response to rising demand can be met more quickly and economically as an alternative to fossil fuel sources.

SMRs require miniaturization of large nuclear reactor components such as fuel support assemblies, spray nozzles, cooling channels, steam separators, and instrument brackets.[14] Some components are difficult to produce with conventional manufacturing, raising capital costs to more than $1 billion and construction time to 4.5 years.[15] AM stands as a means to cut 6 months off this schedule at a 20% lower cost.[16] Maintenance costs also decline as parts are more easily replaced. However, several SMR parts are too large for effective production with current AM rates. The faster rates possible with the EBEAM system can meet this need, but much microstructure and mechanical property characterization remains to be done. Effects of raster speed, beam power, powder size, and porosity from incomplete sintering all need to be established. This work is an effort to answer these questions for 316L stainless steel powder, as commissioned by the DoE under the SBIR program, Phase II (DE-SC0017685).

Experimental Setup

Additive manufacturing was conducted using 316L stainless steel secured from LPW Technology as the raw feedstock. The powder consisted of spherical, gas atomized particles with a diameter of 44–106 μm and an average Hall flow rate of 15s/50g. The powder bed was held under vacuum at $\approx 10^{-7}$ mbar while the beam was magnetically rastered in a snake pattern, tracing in a single direction then turning around at the ends to begin the next row in the opposite direction. This rastering pattern leads to the linear ridging and troughs are apparent in the resulting solid coupons shown below. The EBEAM operating parameters are given in Table 1. Characterization included Rockwell hardness testing, optical microscopy (OM) and scanning electron microscopy (SEM).

Table 1. Electron Beam Processing Parameters for 316L Powder in EBEAM System.

Fixed Parameters	
Beam Energy (MeV)	1.16 ± 0.10
Beam Spot Diameter (mm)	3 ± 1
Powder Packing Fraction	57%
Varied Parameters	
Emitted Gun Current (mA)	1.5–2.25
Collected Beam Current (mA)	0.12–0.18
Beam Scan Rate (mm/s)	0.2–20
Raster Line Overlap	0–50%
Powder Layer Thickness (mm)	1–3

Results

Macroscale material morphology after EBEAM processing is shown in Figure 3. This fabricated from 316L stainless steel powder showing fusion of a rectangular area of material into a single piece is possible. However, the coupon is riddled with hole and ridges from uneven heating and cooling effects. Beam energy and spot size fluctuations (8.6% and 33%, respectively) are the most likely culprit. Efforts are ongoing to improve the beam control to tighten the spread. Another consideration is 316L stainless steel is slightly paramagnetic.[17,18] The local magnetic and electric fields generated from the electron beam is quite large given the large electron velocities, causing a Lorentzian force on the molten pool. This is not verifiable at present as these forces are tied to the beam fluctuations. As beam control improves, the magnetic effects will become consistent.

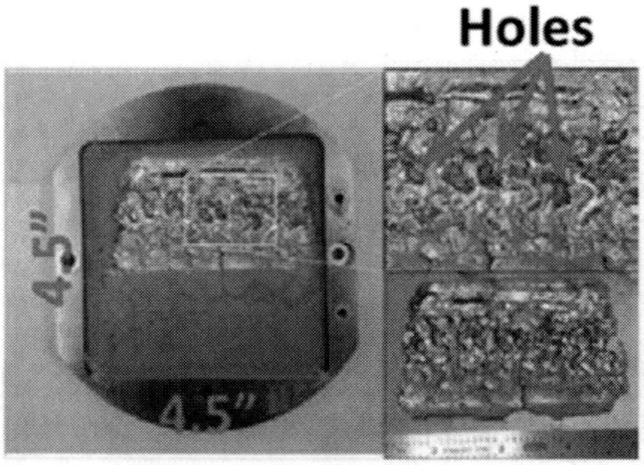

Figure 3. EBEAM test coupon fabricated out stainless steel powder. While forming a complete, solid piece, many holes and ridges exist from uneven heating and wide beam power variation. Efforts are underway to reduce such variation in process conditions.

This uneven morphology has prevented any tensile tests for direct yield strength comparison to the conventional hot-rolled, annealed and pickled (HRAP) 316L stainless steel cladding. Improved control of the process is necessary, including tighter control of the electron beam profile and energy, raster speed, pattern, and dwell time. Figure 4 shows that tightening the beam spot size variation and energy has improved the uniformity in powder melting, eliminating the holes and reducing height variation. Much work remains to be done on raster parameter optimization to eliminate the height variation. Once flat enough samples are fabricated with the AM process, tensile test samples can be cut out for a more direct comparison.

Figure 4. SBIR Phase I coupon vs. the improved version in Phase II. Improvements to electron gun control and tighter energy dispersion shows no holes and smaller ridges, indicating more uniform heating and cooling.

On the other hand, properties such as porosity, hardness, density and grain size in the test coupons can be compared to conventional HRAP plate. Microstructures as seen in OM and SEM are shown in Figure 5. Images of the HRAP plate microstructure in Figure 5a, c are consistent with equiaxial austenitic grains common to 316L.[19] Measured properties after both processing methods are given in Table 2.

Grain sizes in the AM processed material were 200-700 nm, less than 15% of the HRAP grain size. Significant porosity and incomplete sintering was present in the AM coupons, resulting in a low density, only 57% of the HRAP material. Local hardness was still measurable on the AM coupon in well-sintered regions. Hardness values were comparable for both methods, but the AM method is not useful until the porosity is eliminated. As with the large scale mounds and valleys, evening out the melting pool and thermal profile will make for more uniform solidification, reducing porosity.

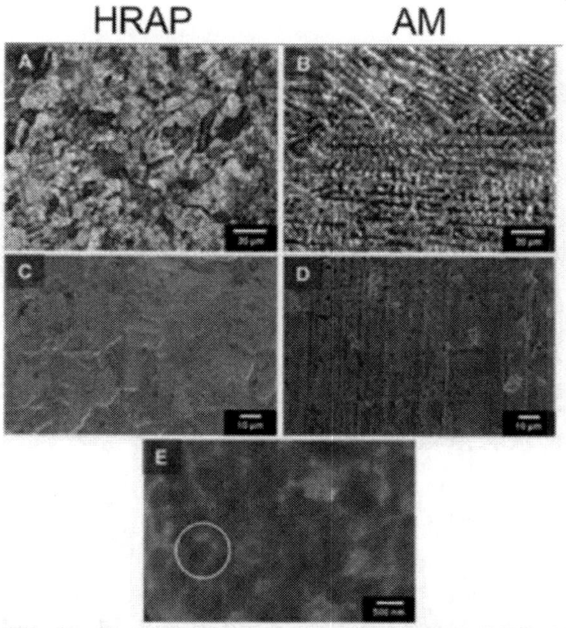

Figure 5. Microstructure after conventional HRAP processing in 316L stainless steel (a, c) and EBEAM powder fusion (b, d, e). The circle in (e) notes a potential grain in the EBEAM-processed material. HRAP produced a uniformly equiaxed structure while EBEAM processing produced a mix of equiaxed and dendritic grains.

Despite the clear uniformity problems in the AM coupons, the technique is still quite promising. This approach is applicable to any part shape, whereas HRAP processing is limited to plate and cylindrical shapes. The rapid heating and cooling rates produce smaller nanocrystalline grain sizes than HRAP plate, resulting in higher strength and hardness. Refinement of the EBEAM approach will reduce porosity and void volume to match or exceed HRAP part properties.

Table 2. 316L stainless steel properties comparison after EBEAM vs. HRAP processing.

	AM 316L Sample Coupon	HRAP 316L Plate	AM vs. HRAP
Hardness (RHC)	33.4 ± 2.5	33.7 ± 0.9	~100%
Density (g/cm³)	4.6	8.0	57 %
Grain Structure	Dendritic + Equiaxial grain structure	Equiaxial austenite	—
Avg. Grain Size (μm)	0.2–0.7	4.8	< 15 %

EBEAM Simulation

Experimental test coupon production is instructive but suffers from limited throughput and high production cost. A large amount of liquid helium is necessary to maintain the accelerator cavity at superconductive temperatures, limiting the number of runs. Computational methods are necessary to weed out undesirable electron beam conditions. To this end, Monte Carlo simulations have been carried out to track electron motion within 316L stainless steel, as shown in Figure 6. From this, the melt pool is predicted along with the temperature profile tail following the rastering direction. Under the beam conditions of 1.16 MeV accelerating voltage, 3 mm beam diameter, 0.12 mA collected current, and 1.9 mm/s raster scan rate, we expected the melt pool to be hemi-ovoidal in shape with 0.82 mm diameter at the plate surface and 0.59 mm depth (Figure 7). Along with incorporating latent heat of melting into the model, we accounted for the fact that 316L does not have a single melting point but rather solidus and liquidus transition temperatures. This is because 316L is an alloy, and therefore, individual grains have their own melting points depending on the local, elemental grain composition. These simulations are helpful for estimating the effects of melt pool size and depth on the cooling time and thus material tensions that result in ridges and uneven height. On the other hand, reducing the melt pool size reduces processing speed. Hence, a balance exists for melt pool size, and work is ongoing to balance experimental beam parameters to obtain a fast speed while keeping height and width variations within acceptable limits requiring a minimum of post-AM machining. Still, a simple model of the melt pool guides powder layer thickness for each spread. A bigger melt pool enables thicker layers faster layer completion, substantially reducing part production time.

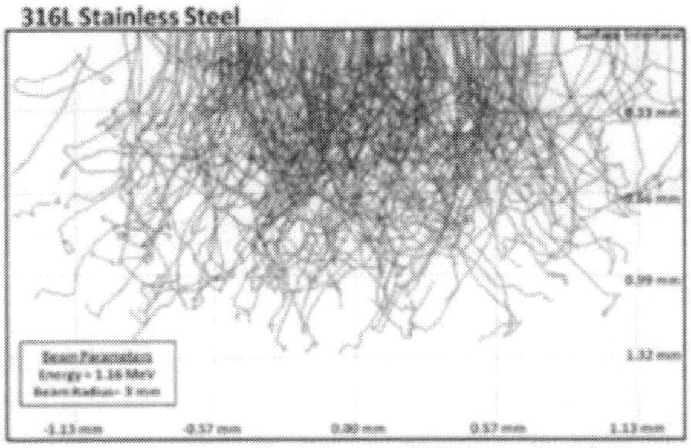

Figure 6. Monte Carlo prediction of electron paths within 316L stainless steel. Blue lines are electron paths, red lines represent electrons reflected back out of the simulated coupon.

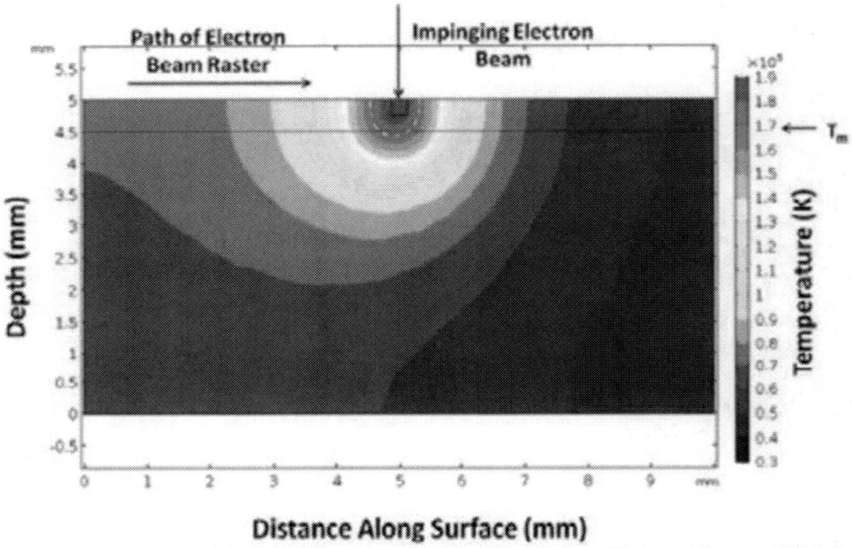

Figure 7. Melt pool prediction from the Monte Carlo simulation. Melt pool size, depth and the cooling of the tail predict heating and cooling rates. This in turn informs how to reduce residual stresses and liquid metal migration into ridges. This also shows the need for low variation in beam parameters.

Conclusions

Mainstream has commissioned an electron accelerator additive manufacturing system that enables many new applications in additive manufacturing. The two-order of magnitude higher electron beam power enables larger, deeper melt pools with 10^6 K/s heating rate and 10^4 K/s cooling rate. Larger parts, faster process speeds and stronger nanocrystalline structures all become possible under these conditions. This equipment was applied to the problem of making larger scale (multiple centimeter) parts for SMRs for upcoming nuclear power applications. 316L stainless steel powder was fused into test coupons and compared to conventional manufacturing methods. The AM coupons exhibited high porosity but similar hardness as conventional manufacturing. AM coupons were riddled with holes, but these have since disappeared with improved control. We expect the ridge height variation to diminish further with improved beam control techniques, leading to a viable large-scale 3D AM process. Once perfected, this technique has many profound implications beyond AM including applying corrosion resistant claddings to irregular parts, leveraging nanograin structure for higher strength parts, ceramic AM and laminar parts of strongly-bonded, alternating layers of material.

Acknowledgements

This material is based upon work supported by the U.S. Department of Energy, Office of Science, Office of Nuclear Energy, under Award Number(s) DE-SC0017685.

This report was prepared as an account of work sponsored by an agency of the United States Government. Neither the United States Government nor any agency thereof, nor any of their employees, makes any warranty, express or implied, or assumes any legal liability or responsibility for the accuracy, completeness, or usefulness of any information, apparatus, product, or process disclosed, or represents that its use would not infringe privately owned rights. Reference herein to any specific commercial product, process, or service by trade name, trademark, manufacturer, or otherwise does not necessarily constitute or imply its endorsement, recommendation, or favoring by the United States Government or any agency thereof. The views and opinions of authors expressed herein do not necessarily state or reflect those of the United States Government or any agency thereof.

References

1 Gibson, I., Rosen, D. & Stucker, B. *Additive Manufacturing Technologies: Rapid Prototyping to Direct Digital Manufacturing*. (Springer, 2010).

2 Abe, F., Osakada, K., Shiomi, M., Uematsu, K. & Matsumoto, M. The manufacturing of hard tools from metallic powders by selective laser melting. *J. Mater. Process. Technol.* **111**, 210-213 (2001).

3 Facchini, L. *et al.* Ductility of a Ti-6Al-4V alloy produced by selective laser melting of prealloyed powders. *Rapid Prototyping Journal* **16**, 450-459 (2010).

4 Abramova, M. *et al.* Grain boundary segregation induced strengthening of an ultrafine-grained austenitic stainless steel. *Mater. Lett.* **136**, 349-352 (2014).

5 Huang, H., Wang, Z., Lu, J. & Lu, K. Fatigue behaviors of AISI 316L stainless steel with a gradient nanostructured surface layer. *Acta Mater.* **87**, 150-160 (2015).

6 Muley, S. V., Vidvans, A. N., Chaudhari, G. P. & Udainiya, S. An assessment of ultra fine grained 316L stainless steel for implant applications. *Acta Biomater.* **30**, 408-419 (2016).

7 Ueno, H., Kakihata, K., Kaneko, Y., Hashimoto, S. & Vinogradov, A. Enhanced fatigue properties of nanostructured austenitic SUS 316L stainless steel. *Acta Mater.* **59**, 7060-7069 (2011).

8 Hammond, V. H. Magnesium Nanocomposites: Current Status and Prospects for Army Applications (ARL-TR-5728). (U.S. Army Research Laboratory, Aberdeen Proving Ground, MD, Sept. 2011).

9 Walsh, S. M., Scott, B. R., Jones, T. L., Cho, K. & Wolbert, J. A materials approach in the development of multi-threat warfighter head protection. (U.S. Army Research Laboratory, Aberdeen Proving Ground, MD, Dec. 2008).

10 Chookajorn, T., Murdoch, H. A. & Schuh, C. A. Design of stable nanocrystalline alloys. *Science* **337**, 951-954 (2012).

11 Guo, X., Yang, G., Weng, G. & Zhu, L. Numerical simulation of ballistic performance of bimodal nanostructured metals. *Materials Science and Engineering: A* **630**, 13-26 (2015).

12 Huang, Y. & Langdon, T. G. Advances in ultrafine-grained materials. *Mater. Today* **16**, 85-93 (2013).

13 Tschopp, M. A., Murdoch, H. A., Kesckes, L. J. & Darling, K. A. ''Bulk'' Nanocrystalline Metals: Review of the Current State of the Art and Future Opportunities for Copper and Copper Alloys (ARL-TR-5728). *JOM* **66**, 1000-1019 (2014).

14 Babu, S., Love, L. J., Peter, W. & Dehoff, R. Report on Additive Manufacturing for Large-Scale Metals Workshop (ORNL/TM-2016/202. (Oak Ridge National Laboratory (ORNL), Oak Ridge, TN, May 2016).

15 Modro, S. *et al.* Multi-application small light water reactor final report. (Idaho National Engineering and Environmental Laboratory, Idaho Falls, ID ..., 2003).

16 NEUP. Fiscal Year 2017 Consolidated Innovative Nuclear Research - Appendix A: Workscopes for U.S. University-led Program and/or Mission Supporting R&D Projects. (Nuclear Energy University Program with

U.S. Department of Energy, Idaho Falls, ID, 2017).

17 Menéndez, E. *et al.* Controlled generation of ferromagnetic martensite from paramagnetic austenite in AISI 316L austenitic stainless steel. *J. Mater. Res.* **24**, 565-573 (2009).

18 AK_Steel. 316/316L Stainless Steel Product Data Sheet. (AK Steel Corporation, West Chester, OH, 2007).

19 Weidmann, E., Guesnier, A. & Taylor, B. Application Note: Metallographic preparation of stainless steel. (Struers A/S, Ballerup, Denmark, February 2016).

577916 Characterization of Spatial Variability in Material Properties in Echogenically-Segmented Metal Components Fabricated By Additive Manufacturing

Background, Motivation, and Objective

The difference in thermal history during layer-by-layer manufacturing of metal components by selective laser melting (SLM) and related additive techniques may result in microstructural anisotropy and spatial variability in strength, density, porosity, and other material properties. The difference is particularly significant between interior and surface regions and in the build direction. While the microstructural variation may be characterized noninvasively by X-ray computed tomography, the options for nondestructive assessment of spatial heterogeneity in end-use material properties are lacking. This paper describes our progress towards the development of a novel approach to the ultrasonic characterization of spatially varying material properties and its application in additive manufacturing.

Statement of Contribution/Method

The traditional nondestructive ultrasonic evaluation gives an aggregate characterization of mechanical properties of solids and is insensitive to their spatial variabilities. This paper adapts the previously developed method for the ultrasonic measurements of segmental temperature distributions in solids [1–2] to estimate the spatial variability of mechanical properties in SLM-manufactured parts. We build upon the earlier report on assessing heterogeneity in additively manufactured aluminum parts [3] and describe our progress in designing echogenic features segmenting 3D printed parts without interfering with end-use functionality while improving the accuracy in segmental characterization.

Results

The application of the developed approach is demonstrated by characterizing segmental variability in the titanium-alloy and stainless-steel parts fabricated by SLM. Three echogenic features incorporated during the fabrication produce the train of echoes, the time-of-flight of which provides information which we use to assess the segmental variability in material properties. The ultrasonic results were found to correlate with the destructive characterization of mechanical properties. Overall, the experimental results indicate the feasibility of detecting heterogeneity of the material properties in 3D-printed parts following the developed approach.

[1] Y. Jia & M. Skliar, *Energy & Fuels*, 2016. [2] Y. Jia et al., *Ultrasonics*, 2016. [3] M. Roy, et. al., *2018 IEEE IUS*, Kobe, Japan.

575361 Lab Enablement through 3D Printing - Nitrogen Inertion Funnel

Hydrogenation reactions pose a significant fire hazard due to the use of a mix of metal catalysts, reagents and solvents. Solvent flames are especially dangerous as they are barely visible and as such a present danger for burn injuries. Most metal catalysts supported on finely divided carbon powder are highly flammable and can ignite solvents and hydrogen. The catalyst material is especially dangerous after it has been used in a hydrogenation reaction leaving the catalyst metal soaked with adsorbed hydrogen. An inertion device has been developed that reduces fire risks associated with the filtering of metal catalysts. The inertion setup is composed of a 3D printed inertion ring that clips onto commercially available filter funnels. This has been tested at two different sized filter applications to determine how much inert gas, nitrogen in this case, was required to reduce the atmospheric oxygen below 12vol% ignition level and has also been tested to see at what flow rates actual ignition is suppressed when methanol is the solvent.

3D printing was a significant factor in not only the design and prototyping of the device but also in the manufacturing of the end use devices. The final product is topologically complex. Most methods of traditional manufacturing would have added significantly to the cost and complexity. This device required non linear internal passage ways and guide vanes to direct and smooth the flow of air. To produce these devices using more traditional means would have necessitated either manufacturing in multiple parts and heat welding or combining many methods, such as sheet metal forming, machining and welding which would have significantly increased cost and limited the design in ways that could have effected the efficacy of the function.

579335 A Scale-Down and Feedback-Control Based Methodology to Optimize Multi-Step Extraction for Botanical Preparation on the Commercial Scale

A systematical methodology from a commercial scale extraction apparatus to laboratory scale based on scale-down and feedback-control strategy was established to optimize the water-extraction process in the manufacturing of a botanical preparation, named as *Xutong injection* in China, on a commercial scale. The raw material of this preparation is *Salvia Miltiorrhiza.* Firstly, quality by design principle was used to achieve process characterization and process optimization of the water-extraction process in a scale-down of 2 L extractor. Several main quality markers of *Salvia Miltiorrhiza* were evaluated to determine the optimal formulation, including the extraction rate of phenolic acids as the effective components and the extraction rate of polysaccharides as the impurities. Secondly, the operation space of process parameters was optimized using a central composite design with response surface methodology as follows: *Salvia Miltiorrhiza* root length was11 cm, the ratio of solvent to raw material was 10 mL / g and each extraction time was 70-90 min for twice. Then, the optimized parameters were applied to make reproducibility test in a commercial scale of 4,000L extractor. In this phase, through in-line data collection, a feedback control strategy was implemented in the pilot production to achieve process real-time determination and control. Finally, the extracting solution manufactured in the commercial-scale process was compared with pre-existing extracting solution collected in the prior batches, which was verified of higher quality and process efficiency. More than 22% active ingredients were extracted than ever. The proposed methodology in this work, once applied to a commercial botanical preparation, is of significant value for continuous improvement in the aspect of reducing time and cost, decreasing batch-to-batch variability and realizing real time release.

571828 Scale-Independent Micronization: QbD Via a Specific Energy Model

It is well established in micronization theory that specific energy provides a holistic approach to evaluating and understanding the particle size reduction operation, and it can be used to fully characterize process performance and robustness. Utilizing a Fluid Energy spiral jet mill operating at ambient temperature, and nitrogen as the feed/grind gas, three operating parameters (Venturi gas pressure, grind gas pressure, and solid material feed rate) can be actively varied in the micronization process, each of which contribute to the calculated total specific energy. This study examined the application of a specific energy model to develop a scale-independent micronization process for commercial production. To develop the model, 34 development runs were performed across a 4" mill and an 8" mill where the Venturi gas pressure, grind gas pressure, and solid material feed rate were varied to investigate the impact of specific energy on the particle size of the micronized drug substance (i.e. output X_{90}). The data gathered at both scales adhere to the same curve, confirming the scale-independent nature of the specific energy model. With specific energy as the underlying mechanism for particle size reduction, the model was used to determine an appropriate registered range for specific energy to achieve the desired particle size distribution. In conclusion, predictive specific energy models can be implemented to create a scale-independent micronization design space; however, these product-specific models will only be applicable when used to predict performance using mills with similar internal geometries, operating principles and conditions, and processing gas.

573647 A Comparative Study of Surrogate Model Based Control Strategy for a Pharmaceutical Crystallization Process

Abstract

In pharmaceutical manufacturing, crystallization is the most preferred technique in downstream processing to recover crystal products. Crystal quality attributes such as crystal size distribution, purity, and polymorphic form, which are critical for not only following unit operations, but also therapeutic properties of the formulated product, are determined by the operating conditions of crystallizers. Therefore, control of a crystallization process is critically and essentially the control of quality attributes. Traditionally, crystallization processes are operated based on a pre-defined recipe of a process variable such as a temperature trajectory for a cooling crystallization system, and the quality of the crystal product is tested only at the end of the process. Quality by testing (QbT) approach leads to high profit loss since batch-to-batch variations of crystal quality attributes are unavoidable, due to the absence of control actions taken during the process in order to avoid effects of any disturbances available in the system through *e.g.* feedback signals coming from solid-state attributes or supersaturation. However, consistency of a process outcome is highly dependent on a robust process control strategy. In literature, emergence of the latest generation of process analytical technology (PAT) tools, mechanistic understanding and modeling of underlying mechanism of crystallization process and incentives by regulatory mechanism have accelerated the efforts on the development of advanced process control strategies in order to achieve desired product quality, so called quality by control (QbC). State of the art crystallization control strategies can be classified into two main groups as model free/data based and model based. As a model free method, direct nucleation control (DNC) is based on the principle of operating at a predefined set point of crystal number density to dominate crystal growth, while supersaturation control (SSC) as another model free method that aims to keep supersaturation level within metastable zone and constant over the experiment time. While the model free methods are simple and found many applications, they might require some investigations on the determination of a robust set point as well as some efforts on online sensor calibration. Model based control strategies utilize real time crystallization process simulations to predict the effects of inputs and disturbances on the process output, whose performance is strictly dependent on the complexity and accuracy of the implemented model as well as speed of solution time. However, often a simple but reliable control strategy is required in the pharmaceutical industries to achieve a fast and successful transition from laboratory to pilot scale crystallization operation. Therefore, in this work we aim to benefit from the simplicity of data based models, while maintaining the accuracy of the predictions in an acceptable range. To this end, we present a surrogate model based control strategy for a pharmaceutical batch cooling crystallization process. Two surrogate models namely Radial Basis Functions Network (RBF) and Polynomical Chaos Expansion (PCE) are utilized to follow a reference trajectory of a mean crystal size by manipulating and updating the temperature profile during process simulation. Performance of the proposed control strategy based on two surrogate models of RBF and PCE is evaluated in the presence of disturbances in the system such as initial concentration, temperature and seed specifications. Additionally, as a proof of concept, experimental applications of the proposed methods are supplemented.

Acknowledgment

We would like to thank the Danish Council for Independent Research (DFF) for financing the project with grant ID: DFF-6111600077B.

559274 Specifications By Design: Framework for Setting Specifications to Maximize Flexibility and Meet Design Requirements

Traditionally, specifications for API starting materials, intermediates and final API are set based on historical data, process capability and limit of quantitation/detection. The flexibility of the process is then determined based on using specification as constraints. Instead of the sequential approach, there is potential for a better outcome by integrating specification setting with process flexibility and design requirements. In this talk, we propose the framework of "Specifications by Design" where the objective is to set specifications for API starting materials, intermediates and final API to balance process flexibility with specification flexibility while meeting design requirements. We will illustrate the framework with few examples: (1) Setting residual solvent specifications for API to balance the flexibility of a key unit operation in the process and the flexibility of specification while meeting the design requirement of desired product performance (2) Setting a isomer impurity specification to balance specification flexibility and meet business design requirements of yield/API cost.

576015 Real-Time Concentration Monitoring Using a Compact Composite Sensor Array for Small-Scale Pharmaceutical Manufacturing Systems

Active pharmaceutical ingredients (APIs) are traditionally produced through time- and cost-intensive large-scale batch manufacturing.[1] The limitations of this approach have been revealed by the Food and Drug Administration (FDA) to be root causes for drug shortages, a major problem in today's pharmaceutical manufacturing.[2] A potential solution to overcome this threat are small-scale, end-to-end continuous manufacturing platforms that integrate synthesis, purification, and formulation within the same uninterrupted microfluidic reactor network.[3,4] This approach possesses a unique high demand on process analytical technology (PAT), for instance, to monitor and determine the strength of aqueous formulations (solution and suspension) *in situ*and in real-time as a critical quality attribute (CQA) prior to the release from the platforms, which are not comparable to lab-scale or industrial-scale needs.[5] The aim of this study is to demonstrate the application of a novel composite sensor array (UV-vis, NIR, Turbidity, Temperature) as a versatile, reliable, cost-effective, and small-scale PAT to measure accurately the concentration of aqueous pharmaceutical formulations (solution/suspension). Four APIs, warfarin sodium and lidocaine hydrochloride, formulated as solutions and acetaminophen and azithromycin formulated as suspensions were measured in their commercial aqueous forms using the composite sensor array. The results were compared with chemometrics data obtained from a state-of-the-art *in situ*Raman probe to provide a reference for the assessment of real-time monitoring. In addition, the concentrations of all formulations were analyzed by partial least squares (PLS) regression, validated and correlated with offline UV-vis spectrophotometer. Results show PLS regressions of $R^2 = 0.9958$ with a Bias = 0.061, which correlated well with offline measurements. The composite sensor array demonstrates potential as an *in situ*real-time PAT for concentration monitoring (solution and suspensions) in one single compact PAT device, as it is preferred in pharmaceutical manufacturing platforms to maintain small footprints.[3–5]

(1) Lee, S. L.; O'Connor, T. F.; Yang, X.; Cruz, C. N.; Chatterjee, S.; Madurawe, R. D.; Moore, C. M. V.; Yu, L. X.; Woodcock, J. Modernizing Pharmaceutical Manufacturing: From Batch to Continuous Production. *J. Pharm. Innov.***2015**, *10*(3), 191–199.

(2) FDA. Strategic Plan for Preventing and Mitigating Drug Shortages http://www.fda.gov/downloads/Drugs/DrugSafety/DrugShortages/UCM372566.pdf (accessed Oct 28, 2015).

(3) Adamo, A.; Beingessner, R. L.; Behnam, M.; Chen, J.; Jamison, T. F.; Jensen, K. F.; Monbaliu, J.-C. M.; Myerson, A. S.; Revalor, E. M.; Snead, D. R.; Stelzer, T.; Weeranoppanant, N.; Wong, S. Y.; Zhang, P. On-Demand Continuous-Flow Production of Pharmaceuticals in a Compact, Reconfigurable System. *Science (80-.).***2016**, *352*(6281), 61–67.

(4) Zhang, P.; Weeranoppanant, N.; Thomas, D. A.; Tahara, K.; Stelzer, T.; Russell, M. G.; O'Mahony, M.; Myerson, A. S.; Lin, H.; Kelly, L. P.; Jensen, K. F.; Jamison, T. F.; Dai, C.; Cui, Y.; Briggs, N.; Beingessner, R. L.; Adamo, A. Advanced Continuous Flow Platform for On-Demand Pharmaceutical Manufacturing. *Chem. - A Eur. J.***2018**, *24*(11), 2776 –2784.

(5) Stelzer, T.; Wong, S. Y.; Chen, J.; Myerson, A. S. Evaluation of PAT Methods for Potential Application in Small-Scale, Multi-Purpose Pharmaceutical Manufacturing Platforms. *Org. Process Res. Dev.***2016**, *20*(8), 1431–1438.

564477 Process Control Strategy for Synthetic Continuous Manufacturing

Continuous manufacturing enables a smaller manufacturing footprint, improved environmental sustainability, improved control strategy, and flexibility to meet changing demand among other benefits; however, the transition from batch to a continuous mode of operation can be resource intensive in terms of number of parameters that require evaluation for optimal process design and process characterization. These resource requirements can be reduced significantly by augmenting empirical design with integrated systems-based modeling.

A 'Virtual Plant for Synthetic Continuous Manufacturing' was developed on the grounds of a systems-based modelling platform. Model components were integrated to form the end-to-end process model, with which *in silico* studies (sensitivity analysis, uncertainty propagation, etc.) are performed to identify the key process parameters that will affect the critical quality attributes. Here we present the development and testing of an integrated advanced process control strategy enabled by the systems-based model. This approach is envisioned to support regulatory filings for CM processes. Development of a virtual plant for synthetic CM enables us to predict process performance and product quality in a real plant environment.

566877 Industry 4.0: Advanced Bi-Layer Control System for Continuous Pharmaceutical Manufacturing Pilot-Plant

The industry 4.0 standard continuous pharmaceutical manufacturing (CM) integrated with real time monitoring, advanced process control (APC), and a novel data management system is highly desired to consistently produce the better quality product with less time and resources. This modernized integrated system also open the possibilities of achieving Quality by Control (QbC) and real time release (RTR) paradigm [1-2]. However, there are different levels of challenges in implementation of industry 4.0 standard CM including advanced process control system. Therefore, this is still an open area of research.

In this work, a bi-layer advanced control strategy has been developed and implemented into CM pilot-plant using industry standard real time monitoring sensors, data management tool and control platform. In first layer, the CPP's and CQA's are controlled in real time using advanced model predictive control (MPC) system while in second layer, the none-confirming products are diverted in real time in waste to assure the final CQA's of qualified tablet lots. This bi-layer control strategy assures the final product quality, improves the production efficiency and minimizes the need of off line testing.

The critical control variables that have been controlled in first layer using model predictive control (MPC) system are drug concertation, powder level before tablet press, main and pre compression forces, tablet weight and hardness. A novel control strategy for powder level control in a chute placed in between blender and tablet press unit operation of continuous tablet manufacturing process has been developed, implemented and evaluated. A noninvasive technique based on change in electric field concept has been used for real time monitoring of powder level in continuous manufacturing pilot-plant. The sensor has been integrated with a control panel (DeltaV (Emerson)) through relays and charms. The control panel then send the signal (4-20 mA) to control platform where the control strategy, a program to operate the sensor and a program to convert the sensor signal into powder level have been implemented. Subsequently, an advanced model predictive controller for powder level have been implemented into our continuous pharmaceutical manufacturing pilot-plant facility. The APC has been used to control feeder, blender and tablet press as well.

A systematic framework including the methods and tools for real time diversion of tablets have been developed and implemented into continuous pharmaceutical manufacturing process as a second layer assurance of product quality. In CM, the drug concentration is measured in real time before the tablet compaction (chute &/or feed frame) using PAT sensor. The proposed control strategy then uses this inlet concentration to determine a signal for the diversion strategy that can accurately be used to reject tablets that are out of tolerance limits at the outlet of the tablet press. Two strategies, i.e. 'fixed window based strategy' and RTD based strategy have been developed, compared and evaluated. In fixed window approach, the tablet diversion is facilitated through knowledge of time delays from the point of detection to the point of the affect (tablet press outlet gate) in the system. The sensor that detects the concentration is connected to a comparator block which decides if the said concentration is within the specifications. If it is not within specification, the experimentally derived time delay is applied and post this the diversion begins. The diversion stops when a concentration within spec is detected and the another time delay is applied. In RTD based approach, the RTD is used to predict the outlet concentration from the inlet concentration. The predicted signal is then used to initiate the diversion. The first approach is simpler to implement but may lead to lower production efficiency. The second approach is based on more advanced technique and will ensure more efficiency but is relatively complex to implement. All the relevant data generated during continuous manufacturing has been systematically collected, stored and organized in a data hub (OSI PI) and cloud system as per industry 4.0 standard.

The objective of this presentation is to demonstrate the performance of integrated MPC and RTD based bi-layer control strategy together with advanced data management system implemented into our continuous pharmaceutical manufacturing pilot-plant facility.

References

1. Bhaskar, A., Barros, F. N., Singh, R. (2017). Development and implementation of an advanced model predictive control system into continuous pharmaceutical tablet compaction process. International Journal of Pharmaceutics, 534 (1-2), 159-178.
2. Bhaskar, A., Singh, R. (2018). Residence time distribution (RTD) based control system for continuous pharmaceutical manufacturing process. Journal of Pharmaceutical Innovation. DOI: 10.1007/s12247-018-9356-7.

562653 A Comparison of Two Bioconversion Processes for the Production of Commercially Important C_4 Streams

Four carbon compounds like 1,4-butanediol (1,4-BDO), succinic acid (SA) have been identified as platform chemicals for producing polymers from renewable resources. The successful commercialization of 1,3-propanediol by Dupont Tate & Lyle Bioproducts and 1,4-BDO by Genomatica Inc. has established the commercial viability of green chemicals. In this talk we present the techno-economic analyses to produce four-carbon compounds (C_4s) like 1,4-BDO, gamma butyrolactone (GBL), and tetrahydrofuran (THF) from dextrose through two separate routes - (i) single step bioconversion and (ii) hydrogenation of bio-based succinic acid (SA). In the single step bioconversion process for production of 1,4-BDO, capital cost of the bioconversion section was found to contribute to around 58 % of the installed equipment cost. Hydraulic load from bioconversion section and membrane filtration was found to contribute the most to operating cost. The process was found to be economically feasible with the minimum sales price (MSP) of 1,4-BDO of $1.82/kg as compared to a market price of $2.5/kg. A mechanical vapor recompression scheme used in the evaporator section was found to benefit the process, increasing energy efficiency and cost effectiveness. In comparison, the SA hydrogenation process, called the C_4s refinery, was not economically competitive given the current state of technological progress. The C_4s refinery produces 1,4-BDO, GBL, THF and ammonium sulfate (AS) fertilizer. The capital costs of the C_4s refinery were found to be 1.8 times higher while the operating costs were found to be 2.5 times higher. An important parameter is the SA fermentation titer. Increase in SA titer values to the maximum values reported in literature were not able to reduce the price of C_4s to below the market price, demonstrating the need for progress in this area. It was concluded that for an extended process involving hydrogenation of SA, additional increases in SA titer values and productivity by genetically modifying the relevant organisms will be required to achieve viability especially in comparison to the single step fermentation process for 1,4-BDO. A secondary reason for unfavorable economics was found to be the dependence of the C_4s refinery on thermal separations which contribute to significant operating and capital costs, technological progress in separation processes such as membranes for organic water separations is therefore necessary. Overall, given the published state of technology, the direct fermentation route produces C_4s (1,4-BDO) at $ 1.82/Kg and is economically viable while the C_4s refinery produces C_4s at $ 3.67/Kg and is not competitive with current production routes.

572716 An Integrated Tool for Rapid Assessment of Chemical Manufacturing Process Emissions, Treatment and Sustainability Performance

The design and assessment of chemical manufacturing processes have been aided by a variety of tools that enable the comprehensive analysis and performance testing of both the steady and transient responses of processes with different levels of complexity. However, comprehensive tools that encompass the evaluation of process emissions, treatment simulation, and sustainability assessment of chemical plant performance are yet to be presented.

In this contribution, the tools developed by the co-authors for simulating Pollution Control Units (PCUs) [1] and performing sustainability assessment with GREENSCOPE [2] are combined for the first time. The PCUs allow for the simulation of several unit operations associated with emission treatment of gas, liquid and solid streams [1], while GREENSCOPE enables sustainability analysis considering four categories: Environmental, Mass, Energy, and Economic Efficiency. Each category has several unique metrics called indicators which provide insight into the sustainability performance of any chemical process. The new integrated tool thus allows for the analysis of scenarios including the impact of current and future models of unit operations on process economics, environmental sustainability, and emissions.

This work therefore presents a unified software framework that provides the opportunity to simulate and perform emission control analysis, pollution control unit sizing and sustainability assessment that could be used to support risk evaluation for different chemical manufacturing scenarios. The developed tool is illustrated via appropriate sections of a biomass/coal gasification process for syngas production with the end goal of acetic acid manufacturing.

References

1. Li S, Feliachi Y, Agbleze S, Ruiz-Mercado GJ ,Smith RL, Meyer DE, Gonzalez MA, Lima FV. A process systems framework for rapid generation of life cycle inventories for pollution control and sustainability evaluation. Clean Technol. Envir., 20(7):1543-1561, 2018.
2. Ruiz-Mercado GJ, Gonzalez MA, Smith RL. Sustainability indicators for chemical processes: III. biodiesel case study. Ind. Eng. Chem. Res. 2013;52(20):6747–60.

577369 Exploring the Potential of Chemical Looping Systems to Provide Alternative H_2 Production Systems Solutions Using Process Simulations and Techno-Economic Analysis

The iron-based moving bed based chemical looping technology developed at the Ohio State University (OSU) is a promising candidate to yield cost benefits in natural gas utilization for a carbon-constrained scenario. OSU has developed four commercially relevant, distinct approaches for high efficiency H_2 production from natural gas[1,2,3]. The first approach focuses on a three-reactor scheme that produces CO_2 and H_2 in separate reactors, providing process intensification benefits[1]. The second approach utilizes a membrane based separation scheme to surpass the thermodynamic limits associated with the H_2 production system in the first approach[2]. The third approach retrofits the methane to syngas based co-current reaction system in a steam methane reformer and considers co-feed of steam and CO_2 for high efficiency H_2 production[3]. A fourth approach focuses on modifying the first three reactor approach and producing syngas from the reducer in addition to the Hydrogen from the oxidizer. This presentation will initially examine, how the application of these four chemical looping systems can lead to process intensification and integration opportunities. The thermodynamic performance will be further quantified by a cold-gas efficiency and an energy efficiency comparison between the three approaches and a baseline Steam Methane Reforming case with 90% Carbon capture. Relevant experiments that include high-pressure kinetic data and sub-pilot scale experimental demonstration will be used as an input for a base techno-economic analysis model for the four approaches. The integrated process and techno-economic analysis conducted with guidance from industrial partners useful and surprising insights into how chemical looping Hydrogen production approaches provide solutions under different adoption and carbon-capture scenarios.

References

1. Kathe, M. V., Empfield, A., Na, J., Blair, E., & Fan, L. S. (2016). Hydrogen production from natural gas using an iron-based chemical looping technology: thermodynamic simulations and process system analysis. *Applied Energy*, *165*, 183-201.
2. Nadgouda, S. G., Kathe, M. V., & Fan, L. S. (2017). Cold gas efficiency enhancement in a chemical looping combustion system using staged H 2 separation approach. *International Journal of Hydrogen Energy*,Â *42*(8), 4751-4763.
3. Kathe, M., Empfield, A., Sandvik, P., Fryer, C., Zhang, Y., Blair, E., & Fan, L. S. (2017). Utilization of CO_2 as a partial substitute for methane feedstock in chemical looping methaneâ€"steam redox processes for syngas production. *Energy & Environmental Science*.

570829 Economic Analysis of Battery (NMC) Cathode Material Production in Flame Spray Process with Sustainable Solvents

The cost of cathode materials contributes approximately 32.7% to the cell construction cost of lithium-ion batteries, which has a significant impact on the price of battery packs. The synthetic routes of cathode material ($LiNi_{1/3}Mn_{1/3}Co_{1/3}O_2$ (NMC333)) based on the solid-state method or co-precipitation method involve complex mixing, drying, and sintering processes, resulting in a material cost too high to reach the U.S. Department of Energy battery price target of \$125/kWh. To reduce the cathode material manufacturing cost, our lab has developed a flame spray process that utilize sustainable solvent glycerol as solvent to manufacture the cathode material. The economic analysis of the new green chemical process was studied based on discounted cash flow method. The major economic indicator used in this study is the minimum cathode material selling price (MCSP), whereas the co-precipitation method for NMC333 production was selected as the reference for economic analysis. Results show that compared with the co-precipitation method, the new flame spray process has the potential to reduce the operation cost by 31% with comparable capital investment and material cost. Moreover, MCSP of the new process is \$19.1/kg, which is 83% of that of the co-precipitation method. When the flame spray process is integrated with the sintering process inline, the MCSP is as low as 15.6 \$/kg, which demonstrates that the new process is an attractive manufacturing process for cathode material production. Furthermore, sensitivity analysis and uncertainty analysis were used to examine the effects of plant scale, raw material price, and analysis parameters on the MCSP and evaluate the certainty interval of the MCSP of the flame spray process.

574815 Soybean Oil As a Bio-Solvent to Sweeten Sour Natural Gas

Soybean oils are readily available bioresources composed of unsaturated fatty acids capable of binding sulfur compounds. Predictive models offer an opportunity to develop cost-effective methods using bio-based solvents to sweeten natural gas. While increases in fracking have increased availability of natural gas – a fuel that when combusted produces less greenhouse gases than coal or gasoline – natural gas is rich in hydrogen sulfide, rendering it sour gas. Hydrogen sulfide is corrosive to processing equipment and poses risks to human health and the environment. Predictive models such as COSMO-RS, the conductor-like screening model for real solvents, can be used to examine interactions at a molecular level, screen solvents, and choose process parameters. COSMO-RS was used to simulate the partitioning of hydrogen sulfide between soybean oil, high oleic soybean oil, free fatty acids, and gas phases. Predicted partition coefficient (K) values ranged from 0.16 to 0.70 for soybean and high oleic soybean oils at temperatures from 10 - 100°C; K values between 0 – 1 are indicative of hydrogen sulfide's preference for the liquid phase over the gas phase. Experimental results demonstrate the feasibility of removing H_2S from gas using soybean oils and are compared to the predicted partitioning. This study provides proof-of-concept for using high oleic soybean oil as a bio-based solvent as a new method for cleaning sour gas.

574828 Reducing Energy, Solvent, and Resin Demand of Sorption-Based Processes for Carboxylic Acid Recovery

Current methods such as ion exchange and distillation to separate bio-based carboxylic acids are material and energy intensive, typically generating waste streams and requiring energy intensive distillation steps to recover a pure product. Sorption-based approaches increase the sustainability of carboxylic acid separations by reducing downstream waste handling, however, current sorption-based technologies typically require large volumes of elution solvent and still result in a dilute carboxylic acid stream. Thus, to obtain a concentrated product, excess water and solvent must be removed from the elution stream via evaporation or distillation which adds a substantial operational cost to the process. This presentation will discuss recent work towards reducing solvent, adsorbent requirements and reducing downstream distillation energy demands in these processes. Specifically, four commercially available weak-base resins were screened by determining their adsorption capacity for butyric acid, a fermentable acid targeted for biofuels. Then, to recover the adsorbed acid, non-aqueous solvents were tested and their regeneration constant were determined. Findings indicate that specific properties of the resins and eluents correlate with the energy, resin, and solvent demands of the system. These results provide a general framework for optimizing sorption-based processes for recovering dilute bioproducts.

572591 Storage of Excess Wind Energy and Utilization of Waste Flue Gas CO_2 through Sustainable Methanol Production

As a larger production of wind energy and other types of renewable energies are predicted in the future, utilizing the excess energy from these sustainable energy sources can increase the energy efficiency of the wind farm by making good use of all the electricity produced. Moreover, CO2 emissions from industry, that contribute to climate change, are also an increasing problem that requires a solution. In this study, we present a novel concept, which aims at solving these two problems by integrating excess energy from wind farms and CO2 from flue gas into an innovative production process using methanol as an energy storage/carrier. The excess wind energy is used to produce hydrogen through water electrolysis and CO2 is captured from flue gas, e.g. from the cement industry. Compared with the traditional methanol production which heavily relies on natural gas as its source of hydrogen, carbon, and energy, this process has a smaller environmental footprint. The novelty of our concept relies on shifting the focus from operating a methanol production plant towards valorizing the excess renewable energy, which is wasted otherwise and considering methanol as a value-added product (as opposed to Szima et al. 2018). Therefore, the plant capacity of 30 000 metric tons per year is scaled based on the quantity of the excess wind energy of the largest wind farm in Denmark. This capacity is significantly lower than those of traditional methanol production plants, which range up to 1.9 million metric tons per year (Ferrostaal, 2015). This adds a constraint to improve the process integration efficiency of the proposed concept. Hence, a techno-economic analysis is performed to investigate the overall feasibility of the proposed concept by identifying key design decisions and essential cost drivers. As results obtained show high sensitivity to both electricity and methanol prices, an uncertainty analysis was conducted in order to determine their effects on the profitability metrics of the proposed novel process. A sustainability analysis is complementarily carried out to emphasize the added environmental benefits of the proposed concept in comparison to the traditional methanol production process.

Reference:

Szima, Szabolcs, and Calin-Cristian Cormos. "Improving methanol synthesis from carbon-free H2 and captured CO2: A techno-economic and environmental evaluation." *Journal of CO2 Utilization* 24 (2018): 555-563.

Ferrostaal. “M5000 Methanol Plant” from *www.ferrostaal.com/en/top-references-ferrostaal/m5000-methanol-plant/* (2015), visited 12.04.2019.

575320 Towards a Novel Energy Financial Security: The Texas A & M Energy Spot Price Index

The available financial instruments providing exposure for the investors to the energy markets are limited to equity based stocks (i.e. S&P 500 Energy) and futures on individual energy commodities (i.e. WTI Crude Oil, Henry Hub Natural Gas). Furthermore, these options are predominantly associated with the oil and gas sector, with almost zero emphasis placed on the renewable energy. Given the lack of a financial security to effectively capture the entire U.S. energy landscape and represent the average cost of energy, we introduce the Texas A&M Energy Spot Price Index (ESPI) as a novel framework to fulfill these needs.

The energy landscape of the U.S. is carefully depicted to determine the products that are directed to the end-use sectors of the U.S. economy [1-5]. The total energy demand of these products, together with their spot prices in the market, serve as the backbone of the Energy Spot Price Index. Since the weights of the Energy Spot Price Index constituents have a lag between two and three months, we introduce a rolling horizon based parameter estimation model to determine the current weights. The spot prices of the products are taken from Bloomberg when available. For products whose spot prices are not available, we introduce and examine three different approaches [6-8]. The methodology is illustrated on 120 months to show the capability of accurately determining the value of the Energy Spot Price Index. The proposed Energy Spot Price Index is the first of its kind that precisely quantifies the average market price of energy. As such, it can be utilized as an efficient tool for all investors who look for exposure over the average market price of energy, trading over the whole energy market, hedging their investments, as well as a benchmark for measuring the overall cost of energy. Therefore, the Energy Spot Price Index can be used as a tradable financial security (e.g. ETF) that is updated daily providing an alternative for the investors who look to diversify their portfolios.

References:

- Elia, J. A.; Baliban, R. C.; Xiao, X.; Floudas, C. A. Optimal energy supply network determination and life cycle analysis for hybrid coal, biomass and natural gas to liquid (CBGTL) plants using carbon-based hydrogen production. Computers and Chemical Engineering 2011, 35, 1399-1430.
- Elia, J. A.; Baliban, R. C.; Floudas, C. A. Nationwide energy supply chain analysis for hybrid feedstock processes with significant CO2 emissions reduction. AIChE Journal 2012, 58 (7), 2142-2154.
- Elia, J. A.; Floudas, C. A. Energy Supply Chain Optimization of Hybrid Feedstock Processes: A Review. Annual Review of Chemical and Biomolecular Engineering 2014, 5 (1), 147-179.
- Onel, O.; Niziolek, A. M.; Floudas, C. A. Integrated biomass and fossil fuel systems towards the production of fuels and chemicals: state of the art approaches and future challenges. Current Opinion in Chemical Engineering 2015, 9, 66-74.
- Floudas, C.A.; Niziolek, A. M.; Onel, O.; Matthews, L.R. Multi-Scale Systems Engineering for Energy and the Environment: Challenges and Opportunities. AIChE Journal 2016, 62 (3), 602-623.
- The Texas A&M Energy Price Index: A Predictive Framework, 2019
- Energy Information Administration (EIA), Electric Power Monthly, 2019
- U.S. Department of Energy (DOE), Green Power Markets, 2018

566273 Techno-Economic Analysis of Septic Tank Waste Treatment in a Centralized Hydrothermal Treatment Facility

Hydrothermal treatment, a thermochemical conversion process that is particularly useful for wet wastes, of septage produces valuable solid and liquid fertilizers while eliminating human pathogens. Additionally, it treats a problematic waste that is a burden for most traditional waste water treatment facilities. Operation of a centralized treatment facility would augment an area's capacity to utilize the nutrients and valuable products available in septage while also mitigating its potential environmental hazards. This facility would be alternative to landfilling, which can be detrimental to public waterways, or treatment at a traditional waste water plant. This analysis investigates the most important factors for an economically feasible facility that both treats septic tank waste and produces fertilizer products. The environmental impact of the plant is also compared to the standard methods of septage treatment.

578172 Turning a Hazard into Sustainable Revenue Streams: H_2S Valorization in Offshore Oil and Gas Productions through Process Intensification

The grand challenges of the 21st century lie squarely at the interplay of water, energy, and food— the so-called water-energy-food nexus. Add climate change into the mix, and the challenges become *more* challenging. Aggressive measures to decarbonize the energy market are necessary to tackle the threat of climate change. However, due to the ever-increasing global population, energy requirements are expected to increase 50% by 2030.[i] While ensuring that we protect the natural environment, it is also important to provide energy security. This calls for an energy supply that is consistent, reliable, sufficient, and affordable. Natural gas has arguably the least carbon footprint among fossil energy resources (carbon dioxide emission of burning natural gas is approximately half that of coal). While natural gas cannot be viewed as the permanent solution to energy security, it plays a vital role as the bridge fuel in the transition to a low-carbon economy.

Hydrogen sulfide (H_2S) is a major hazardous by-product in oil and gas production as well as many bio-based manufacturing processes, such as anaerobic digestion. Desulfurization of sour gas is essential prior to transportation or further processing due to its extremely toxic, acidic, and corrosive nature. Several offshore oil and gas production platforms are suffering from elevated levels of H_2S in recent years, and the production facilities are not designed for such high levels of H_2S. Currently, H_2S is removed using a non-regenerative scavenging process. The scavenger reacts with H_2S and produces less acidic, water-soluble compounds in an irreversible reaction.

Although the current process is rather established, it comes with significant challenges. Such challenges include high OPEX due to excessive amounts of the scavenger used in offshore productions, environmental concerns coming from the increasing levels of H_2S and increased mass of discharged spent scavenger, and fouling issues creating blockages in several downstream unit operations, such as wastewater treatment facilities.

For these reasons, design and development of more sustainable approaches that can fulfil the H_2S content threshold limit requirements and that can eliminate the disadvantages of the current non-regenerative H_2S-scavenging process are deemed necessary, specifically from an economic point of view. This work focuses on the design and techno-economic analysis of an intensified process that can be installed on an offshore platform to simultaneously remove and convert H_2S from produced natural gas to value-added product(s) to generate new revenue streams. The added benefits of this new process are that it allows for the efficient recovery of resources by turning the hazard, H_2S, into an opportunity. The proposed process is therefore well suited for the application of intensification concepts, integrating reaction and separation into a single unit.

[i] IEA, "World Energy Outlook 2012," International Energy Agency, 2012.

554760 Critical CO_2 Carnot Engine for Industrial Waste Heat Recovery and Utilization

Industrial processes regularly discard large volumes of low-grade waste heat to the environment. Harnessing this heat and transforming it into usable power would be environmentally and economically advantageous for industries. An overlooked solution to utilizing this waste heat is the use of a Carnot cycle to generate work with supercritical CO_2 as the working fluid. Supercritical $CO_2$is a novel working fluid for a Carnot engine as its critical temperature (31.1C) is much lower than the boiling point of water. This allows for the lower isotherm of the Carnot cycle to be constructed at the critical temperature, resulting in a sizable Carnot cycle with a near flat lower isotherm and upper isotherm just below the temperature of the low-grade heat. Thus, this type of Carnot cycle encompasses a larger area on a P-V diagram than any other thermodynamic conditions, thereby optimizing the amount of work that can be converted to power using a relatively small engine.We describe this approach of waste heat harnessing and illustrate ideas for utilizing the waste heat from an industrial toluene process. The impact of compression ratio, upper isotherm temperature, and pressure in the engine are investigated. Resulting savings amount to hundreds of thousands of kWh/year, thereby reducing cooler loads and electricity bills for the process, as well as reducing environmental emissions.

References:

[1] J.T. Banchero, B.D. Smith (Ed.), R.J. Hengstebeck. *Disproportionation of toluene*, 1969.

[2] Eldred H. Chimowitz Madeleine R. Laitz, F. Douglas Kelley. *Critical CO_2Carnot Cycle for Waste Heat Utilization*, 2017.

INDUSTRIAL SYMBIOSIS IN TRINIDAD AND TOBAGO: A FOCUS ON CARBON DIOXIDE

Thérèse G. Lee Chan and David A. Janes, The University of the West Indies, St. Augustine, Trinidad and Tobago

Abstract

Enterprise input-output (EIO) models are used to analyze the flows of resources, such as energy and materials, around industrial networks. The majority of the greenhouse gas, carbon dioxide (CO_2), emitted in Trinidad and Tobago (T&T) is from industrial sources. In this study, a simple representative model was developed of a network of CO_2 exchanges within an industrial park in T&T. The EIO model was developed from both process engineering and economic data. It incorporated performance indicators, which had been proposed in the literature, to calculate the level of industrial symbiosis (IS) and benefits - based on "the three pillars of sustainability" - realized in the representative industrial network. The eco-connectance, which is one measure of the level of IS, was determined as 1.3 for the representative network, thereby indicating the presence and level of IS in the industrial park. Furthermore, optimized flow networks were created, with and without a CO_2-reusing propylene carbonate plant. Multi-objective optimization was used to determine how to add the CO_2-reusing plant to the industrial network to minimize both CO_2 emissions and implementation costs. The addition of a CO_2-reuse plant reduced the CO_2 emissions by 3%, demonstrating there is scope for improving the existing IS network.

Introduction

Industrial symbiosis (IS) can be regarded as using, recovering and redirecting resource-laden wastes for reuse, in a network of separate commercial operations, so resources remain in productive use for longer. IS is considered an effective approach to waste management (Yazan, Romano et al. 2016) delivering both economic and environmental benefits. The twin-island state of Trinidad and Tobago (T&T) is rich in hydrocarbon resources, and was amongst those countries generating the highest carbon dioxide (CO_2) emissions per capita at 25.72 tonnes per head in 2016 (Knoema World Data Atlas, 2018). In 2014, 74% of the CO_2 emitted in T&T originated from industry (UNDP, 2014). The majority of the petrochemical industry in Trinidad is housed in the Point Lisas Industrial Estate (PLIE), which contains several ammonia and methanol plants.

In this paper, the exchanges of process CO_2 between various petrochemical plants in PLIE are used to assess the level of industrial symbiosis occurring on the industrial park. The aim of this work was to develop a simple representative model of this nascent symbiotic industrial system, in terms of CO_2 exchanges at PLIE. Our model was then used to perform an integrated process engineering and economic analysis to quantify the level of exchanges and benefits beneath "the three pillars of sustainability" and thence determine the scope for improvement.

In part because there are currently no legislative restrictions on CO_2 emissions in T&T, PLIE is not thought of locally as being a bastion of progressive industrial ecology. Although PLIE was not originally planned for industrial symbiosis, waste CO_2-streams have been shared amongst plants since 1983 (Boyle and Baetz, 1998) when the Trinidad and Tobago Urea Company plant was commissioned to use waste process CO_2, ammonia and some facilities from a nearby separately-run Fertrin ammonia plant to produce urea. The current "self-organized" state of sharing of waste streams developed over the years, rather like early sharing at archetypical eco-industrial park (EIP) at Kalundborg, Denmark (Ehrenfeld and Gertler, 1997), in that each additional exchange in the park was an independently negotiated business deal, which was established only if expected to be commercially beneficial.

Few studies have been conducted on the application of industrial symbiosis in T&T. In 1998, four companies in PLIE were evaluated in terms of waste management and the study (Boyle and Baetz 1998) found that there was potential for the waste reduction by reusing and recycling, although the study did not consider CO_2 as a waste resource. In 2008, Boodlal (2008) compiled an inventory of CO_2 resources in Trinidad, in which the emissions from the chemical plants at PLIE were calculated with appropriate stoichiometric ratios for the relevant reactions.

Methodology

There is no standard methodology for measuring the level of IS in an industrial network or park. Some authors use economic, environmental and social indicators to show how IS improves a park by comparing the indicators for the park with and without IS activities. However, this does not give an actual measure of the level of IS. Some of the proposed measures for industrial symbiosis are eco-connectance, by-product and recycling rate (Tiejun 2010), eco-efficiency (Park and Behera 2014), environmental impact momentum (Yilmaz et al., 2016) and technical exchange efficiency (Fraccascia, Albino et al., 2017). Yilmaz et al. (2016) published a comprehensive list of indicators that can be used when considering industrial symbiosis.

Material flow analysis (MFA), enterprise input-output (EIO) tables, life cycle analysis (LCA), eco-efficiency, eco-connectance and waste recovery are some of the tools which have been used in reducing waste in a system in a sustainable manner. MFA, LCA and eco-efficiency are not preferred since they do not consider the whole picture.

In this study, the process CO_2 emissions for selected plants were evaluated based on reported information about the various technologies employed and the plant capacity. Flue-gas CO_2 emissions were excluded from this study, because of the much lower CO_2-purity of these streams. Potential means for improving the sharing and reusing of CO_2 to create shared value, that is the IS level, were considered. Nine petrochemical plants in PLIE were selected as being representative of the industrial park. The plants considered were ammonia, methanol, urea and melamine. An EIO table for the industrial park was developed in MATLAB® (ver. R2019a), the chosen modeling environment.

The general equations and variables for an EIO system are:

$$X = Ax + f \tag{1}$$

$$r = Rx \tag{2}$$

$$w = Wx \tag{3}$$

Variables:

A – Technical coefficient matrix
x – Total output vector
f – Vector of final demands
R – Resource intensity matrix
r – Vector of primary input
W – Waste intensity matrix
w – By-products and waste vector

It is assumed that there is only one main product for each process. A, R and W are matrices of constants and are typically observed or calculated values for a specified technology for the process. Therefore, if process demand is changed, the outputs from the process would be changed by that value.

Three EIO tables were developed for a *base*, *current* and *proposed* case. The base case was one in which there was no symbiosis occurring and this was used as a baseline in the study. In the current case, the existing symbiosis was shown and in the proposed case, a CO_2-reusing plant was added to the industrial park. Material and energy flows for the base and current cases were estimated on the basis of the average production rate of the various plants over a three-year period. Various plants, which reuse CO_2, were considered, such as manufacturing processes for biofuels (Benn-Ogeer et al., 2015), propylene carbonate (Demirel, 2015) and polyethercarbonate polyols (von der Assen and Bardow, 2014). The propylene carbonate plant was selected based on the feasibility and the availability of published performance information. The data used to simulate and cost the plant in Aspen Plus® (ver. 10) was obtained from Demirel (2015). For all the cases, the direct and total coefficients were determined. Besides the evaluation of the direct and total coefficients, the level of industrial symbiosis occurring in the various scenarios was evaluated using various industrial symbiosis measures and sustainability indicators.

In this study, the focus was on the exchange of process CO_2 amongst the plants. The CO_2-rich process streams, which originated from the ammonia plants, were supplied to the sinks at an assumed CO_2-purity of 99 wt%. US dollars were used for all prices and costs. The price of CO_2 supplied to the sinks was determined from the calculation of the levelized cost of CO_2 transportation and an assumed raw material cost for CO_2, to buttress the mutualism of the IS. Consequently, pipelines from the CO_2 sources to sinks were evaluated. The equation developed by McCoy and Rubin (2008) was used to estimate the diameter of the pipelines required for the sink flows. The costs associated with purification and compression of the CO_2 were taken into consideration and together with the diameter and length of the pipeline, these were used in the determination of the levelized cost of CO_2 transportation ($ tonne^{-1} of CO_2).

Optimization of the industrial park was computed in MATLAB by taking into consideration the amount of waste CO_2 and the economics. Single objective optimization for each case was carried out to minimize waste CO_2 and to minimize the cost. Cases were considered using linear programming (LP) and mixed integer linear programming (MILP). In the MILP, binary variables were used to prevent small flow rates through the pipelines. For the LP and MILP cases, the dual-simplex algorithm and the branch and bound algorithm were respectively utilized. Finally, multi-objective optimization using the genetic algorithm in the MATLAB function *gamultiobj* was used to minimize the amount of waste CO_2 and the cost. The Pareto front for each case was generated.

Results and Discussion

Material and energy flows were calculated for the input and output of the plants in the industrial park. These flows were then used to generate the EIO tables for the base, current and proposed cases. The cases showed that from the ammonia plants, there were approximately 202,000 kg/h of nearly pure CO_2 being produced. The EIO for the current case indicated that approximately 35% of the CO_2 was reused in methanol and urea plants in the industrial park and the remainder was vented to the atmosphere. Introducing a new plant, which reused CO_2 from the industrial park, increased the CO_2 usage by approximately 3%, thereby reducing the amount of CO_2 emitted to the atmosphere. From the EIO table, the direct and total coefficients were determined. Analysis of the direct coefficients showed that the largest direct coefficient for the by-products and waste was attributed to the nearly pure CO_2 being emitted from the ammonia plants.

The core EIO model gives an idea of the material and flow exchanges within the network, but does not measure the level of industrial symbiosis. The calculated values for the performance indicators, used to measure IS in these three cases, are shown in Table 1.

Table 1. Industrial Symbiosis Performance Indicators

Industrial Symbiosis Indicator	Base	Current	Proposed
Eco-connectance	0	1.3	1.8
Environmental impact momentum	0	0.56	0.62
Fraction of waste CO_2 utilized	0	0.36	0.38

In the base case, it was assumed that there were no exchanges amongst the plants. The values obtained for the current case show that there is some level of industrial symbiosis occurring and those for the proposed case indicate that the addition of a plant which utilizes CO_2 as a raw material can lead to increased levels of industrial symbiosis.

In this study, a propylene carbonate plant was proposed for the industrial park. There are several ways in which CO_2 can be utilized, but because many of these uses are emerging technologies, it is difficult to obtain information for them. The propylene carbonate plant had a capacity of approximately 10,500 kg/h and consumed 4,500 kg/h of CO_2. This capacity was determined in accordance with the current market for propylene carbonate. The results from the model indicated that the feasibility of such a plant is highly dependent on the price of the catalyst since this was the most expensive component of the model.

One of the objectives of this study was to optimize the flow of CO_2 in the three cases. This optimization considered minimizing the waste CO_2 and the cost. The levelized cost of transporting the CO_2 was calculated and hence pipelines from the CO_2 sources to the sinks were considered. The transportation costs of CO_2 ranged from US$9 per tonne to $50 per tonne of CO_2. The higher end of the costs was attributed to smaller flows in the pipelines and hence due to a fixed cost recovery period, the price would consequently be higher. It should be noted that the costs obtained were derived from a fixed flow in the pipeline and did not take into account the change in the flow rate and the cost. This was done to ensure that the optimization model remained linear. According to literature, there is no fixed cost for CO_2 since it is dependent on the purity of CO_2, distance and the state in which the CO_2 is transported.

Various optimization cases were considered using the current and proposed data. In the first scenario, minimization of waste CO_2 was carried out. MILP and LP were used to solve the optimization models for the base and proposed cases. Results from the LP case with the current data indicated that the CO_2 from four out of the five plants should be sent to the CO_2 sinks. The MILP results indicated the same but the flows were different for the plants. However, for the proposed cases both MILP and LP showed that flows should be sent to the sinks from all the plants. It is suggested that some flows be sent from each plant since the existence of a plant venting all its CO_2 to the atmosphere would increase the environmental burden for that particular plant.

The optimal annual costs for the current and proposed cases were \$12 million per year and \$14 million per year respectively. It is evident that in the current case, the transportation cost for CO_2 is less since there are less sinks which require CO_2. It should also be noted that for the proposed case, there would be some additional expenditure in constructing and maintaining the propylene carbonate plant. Multi-objective optimization for minimizing the CO_2 waste and cost was also done for both cases. The annual costs for each case were \$12.2 million per year and \$14.6 million per year, respectively.

Sustainability indicators were employed to evaluate the optimized cases using multi-objective optimization (see Table 2). The eco-efficiency was determined as a ratio of the annual production to the amount of CO_2 used.

Table 2. Sustainability Indicators

Indicator	Sustainability Dimension	Units	Current Case	Proposed Case
Eco-efficiency	Environmental	-	9.8	9.4
CO_2 emissions	Environmental	t/h	277	273
Resource efficiency	Environmental	t/h	0	4.5
Relative change of CO_2 recycling	Environmental	%	36	38
Total Capital Investment (CO_2 transport)	Economic	\$	22.5	26.5
Total Capital Investment (new plants)	Economic	Million \$	-	17.2
Profit as a result of IS activities	Economic	Million \$/y	0.41	(0.43+3.6)
Job Creation	Social	-	No	Yes
Creation of Industrial Symbiosis	Social	-	6	7

The new plant led to a reduction in the overall CO_2 emissions thereby increasing the amount of CO_2 being reused. Addition of the propylene carbonate plant resulted in a decrease in eco-efficiency. Nevertheless, it would lead to an increase in jobs and would create an increased level of industrial symbiosis in the industrial estate.

Further work may investigate the actual linkages amongst the plants and determine the capacities of the existing pipelines. It could also consider looking at CO_2 in flue gas and other CO_2 utilization options. Additionally, exchanges of other materials and energy from the industrial park need to be included to get a fuller picture of how industrial symbiosis is being and could be put into practice.

Conclusions

The model indicates that current partners in the on-going industrial symbiosis at the Point Lisas Industrial Estate ought to be achieving sustainable benefits. The estimated levelized cost for CO_2 transportation ranged from \$9 per tonne to \$50 per tonne of CO_2. The introduction of a few chemical manufacturing plants reusing CO_2 should lead to increased IS indicator values and further triple-bottom-line benefits, even though it would only make a small dent in the quantity of as yet unused process carbon dioxide. The integration of data from process engineering models into EIO models provides a means to analyze and optimize potential IS scenarios.

References

1. Benn-Ogeer S. K., D. A. Janes and R. Deo (2015). "A Life Cycle Assessment (LCA) of the Production of Biofuels in Trinidad from Waste CO_2 With Microalgae", e-Journal of the Caribbean Academy of Sciences 8(1).

2. Boodlal, D. (2008). "Trinidad and Tobago's CO_2 Inventory and Techno-economic Evaluation of Carbon Capture Options for Emission Mitigation." 3rd Tobago Gas Technology Conference, Lowlands, Tobago.

3. Boyle, C. A. and B. W. Baetz (1998). "A prototype knowledge-based decision support system for industrial waste management: Part II. Application to a Trinidadian industrial estate case study." Waste Management 17(7): 411-428.

4. Demirel, Y. (2015). "Sustainability and Economic Analysis of Propylene Carbonate and Polypropylene Carbonate Production Processes Using CO_2 and Propylene Oxide." Chemical Engineering & Process Technology 0: 6:3.

5. Ehrenfeld, John; Nicholas Gertler (1997). "Industrial Ecology in Practice: The Evolution of Interdependence at Kalundborg". Journal of Industrial Ecology 1: 67–79.

6. Fraccascia, L., V. Albino and C. A. Garavelli (2017). "Technical efficiency measures of industrial symbiosis networks using enterprise input-output analysis." International Journal of Production Economics 183: 273-286.

7. Knoema World Data Atlas. (2018). "Trinidad and Tobago – CO_2 emissions per capita". [retrieved 01/04/2019]. Available from: https://knoema.com/atlas/Trinidad-and-Tobago/CO2-emissions-per-capita.

8. McCoy, S. T. and E. S. Rubin (2008). "An engineering-economic model of pipeline transport of CO_2 with application to carbon capture and storage." International Journal of Greenhouse Gas Control 2(2): 219-229.

9. Park, H. S. and S. K. Behera (2014). "Methodological aspects of applying eco-efficiency indicators to industrial symbiosis networks." Journal of Cleaner Production 64: 478-485.

10. Tiejun, D. (2010). "Two quantitative indices for the planning and evaluation of eco-industrial parks." Resources, Conservation and Recycling 54(7): 442-448.

11. UNDP, Factor CO_2. (2014). "Projection of the Business as Usual (BaU) and measures scenarios of carbon emissions for the electrical energy generation, industry and transportation sectors". Port of Spain, Trinidad and Tobago.

12. von der Assen, N., and A., Bardow, (2014). "Life cycle assessment of polyols for polyurethane production using CO_2 as feedstock: insights from an industrial case study." Green Chemistry 16: 3272-3280.

13. Yazan, D. M., V. A. Romano and V. Albino (2016). "The design of industrial symbiosis: an input–output approach." Journal of Cleaner Production 129: 537-547.

14. Yilmaz, O., E. Yontem and E. Alkaya (2016). Industrial Symbiosis Indicators. FISSAC.

569669 Two-Phase Optimal Design of Pharmaceutical Separation: Sampling-Based Uncertainty Analysis and Reinforcement Learning

Pharmaceutical industry has been consistently growing up because of diverse factors such as the increased size of the global aging population, drug affordability, government policies, and supply-side factors. It has reached for a number of challenges including a changing health care landscape, expiring patents and generic competition, pricing pressures, and a persistent economic slowdown. One of the feasible solutions to overcome the current situation is to develop an optimal and smart separation process of pharmaceuticals in order to improve the performance and maintain the system flexibility under the uncertainty. Liquid-liquid extraction (LLE), which is also known as solvent extraction, is the main process that is much employed during the downstream separation. In the LLE process, a feed stream of two or more components is faced with a second liquid stream containing solvent. Therefore, the use of proper solvents and the implementation of an optimal operating design for the LLE process are inevitably important issues to keep obtaining high yield of pharmaceuticals. This research aims to construct an advanced LLE framework including two-phase strategy. The first phase describes the optimal design of the LLE process based on thermodynamics and transport properties and uncertainty analysis. The second phase presents a smart operating planning by reinforcement learning according to results from the first stage. In the reinforcement learning part, a deep Q-learning algorithm to control the LLE process is investigated. Each stage in the purpose of this study is underpinned by several internal steps as the follows. For the first step in the first stage, electrolyte non-random two-liquid segment activity coefficient (eNRTL-SAC) model is employed to estimate segment parameters and activity coefficients of a target pharmaceutical. eNRTL-SAC model is reminiscent of NRTL-SAC model and additionally uses an electrolyte conceptual segment parameter. Segment parameters and activity coefficients are explicitly utilized to perform solubility modelling. The next step is that liquid-liquid extraction column as a separation process is considered and distribution coefficients are calculated based on results of thermodynamics properties; thereafter, uncertainty analysis is conducted to identify optimal solvents and operating conditions. Reinforcement learning in the second stage technically aims to optimize an operating structure using the results from the first stage. For the first step in the second stage, an agent and an environment have to be precisely defined, each of which corresponds to the controller and the LLE process respectively in this research. The agent in the second step takes an action to the environment whereby the environment operates the designed process model according to the action. The environment in the last step discharges states and rewards to the agent to be updated by deep learning algorithms. Actions, states, and rewards can be constructed in various ways as the relationship in between the agent and the environment should be considered.

Property process modelling and uncertainty analysis

Screening of an optimal solvent or solvents mixture is an inherent procedure to formulate an efficient pharmaceutical process. Solubility modelling is most often used as a criterion of solvent screening and requires a knowledge of a proper thermodynamic model. As aforementioned, eNRTL-SAC model is applied to this study, whereby regressed segments parameters as well as activity coefficients are able to be disclosed. Activity coefficients are determined by regressed segment parameters and a distribution coefficient, which is also known as a partition coefficient and means the ratio of concentration of a compound in a mixture at equilibrium, is calculated by activity coefficients. Extraction factor consists of the distribution coefficient, the flow rate of the feed stream in the LLE process, and the flow rate of the solvent stream in the LLE process, and Kremser-Souders-Brown theoretical stage equation directly uses the distribution coefficient and the extraction factor to estimate the number of stages in the extraction column that ultimately affects the operating costs. Parameter uncertainties and an influence thereof are implemented by uncertainty analysis in order to give a decision maker the guideline of an optimal process design. Sampling-based model solution prior to uncertainty analysis is conducted as means to explore and identify the optimal process variables thereby Monte Carlo method is exploited to illustrate the influence of parameter uncertainties.

Off-policy control with deep Q-leaning

General reinforcement learning basically encompasses five components, each of which is state, state transition probability matrix, action, reward, and discount factor. Action is determined by a policy included in the agent. Conventional objective function in reinforcement learning is to maximize a reward function according to the policy. Therefore, understanding of the optimal policy is the main purpose in reinforcement learning. Off-policy strategy is associated with the concept of exploration and exploitation. In the case of on-policy strategy, a specific policy can be upgraded through agent training, on the other hand, it has the weakness of exploring solutions available in the vicinity of the current policy. At the expense of consuming computational costs, off-policy can flexibly explore distinct candidates and approach much closer to the global solution. The agent is designed by deep learning algorithms to suggest the optimal action every single episode. State information and value of reward from the environment play a role of input variables of the agent and suitable hyperparameters of the deep learning algorithm have to be set up to improve the efficiency of the agent.

This study purposes to design a novel extraction process in pharmaceutical industry based on uncertainty analysis and reinforcement learning. Thermodynamics property modelling, solubility modelling, solvent screening, and uncertainty analysis are consecutively examined to design the feasible LLE process of pharmaceuticals. Afterwards reinforcement learning is considered to suggest the optimal operating planning based on the results from uncertainty analysis. Moreover, the proposed two-phase framework of the LLE process would be extended to other processes in downstream process.

573815 Computer-Aided Design of Solvent Blends for the Cooling and Anti-Solvent Crystallisation of Active Pharmaceutical Ingredients

The majority of pharmaceutical products are delivered in the solid form, such as tablets or via aerosols. As such, crystallisation remains the process of choice for the manufacturing of active pharmaceutical ingredients (APIs) – typically via solvent-based techniques. Thermodynamically, solvent choice will affect the solubility, and thus crystal yield of the API, but the effects of this decision on transport properties, process and product safety, and the final crystal morphology – which can impact downstream processing and ultimately the *in vivo* efficacy of the drug – are all driving forces for selection. For novel drug molecules, solvent selection is often performed via time-consuming and expensive experiments [1], limiting the full range of solvent mixtures and process conditions from being explored. Consequently, fast development of manufacturing processes that achieve high standards in terms of productivity, quality and environmental performance can be difficult.

Over the last decades, Computer Aided Molecular Design (CAMD) has been developed to overcome such difficulties, with the ultimate aim of guiding experiments towards optimal candidate molecules [2,3]. To reduce the number of potential solutions in such designs, a decomposition-based approach has been proposed [4], whereby smaller subproblems are posed and then solved successively. By fixing process conditions and utilising a single crystallisation technique, this has been applied to crystallisation design [5]. Indeed, most existing methodologies focus on the selection of single solvents, utilising a specific crystallisation technique, whereas more integrated problems, in which cooling and anti-solvent effects are combined and solvent mixtures are considered, have not yet received significant attention. This is likely due to the complexity of formulating and resolving a mixed-integer optimisation problem to represent these design choices; such problems result in challenging non-convex design spaces for the continuous variables, in addition to the combinatorial solution space.

More recently, the use of Generalised Disjunctive Programming (GDP) [6] within the Computer Aided Mixture/blend Design (CAM^{b}D) framework has been proposed to design optimal solvent mixtures that maximise solubility of an API [7,8]; optimal mixtures were shown to outperform pure compounds. This strategy can be used to use to solve more integrated problems, enabling the user to explore the relative merits of cooling crystallisation, anti-solvent crystallisation and combined approaches concomitantly and solvent mixtures can be considered throughout [9], but this type of approach has not yet received significant attention. We investigate the formulation and solution of such an integrated crystallisation design problem, whereby the optimal process temperature, selection of solvent and anti-solvent molecules, and design of the solvent mixture composition, can be identified simultaneously – resulting in a mixed integer nonlinear problem (MINLP). Based on an equilibrium model, the formulation considers solvent use and crystal yield, but can be extended to account for other factors. To calculate the relevant thermodynamic properties, we use the SAFT-γ Mie equation of state, which has been shown to provide high quality predictions of solubility [10].

This design formulation is implemented in gPROMS and successfully applied to a case study on the crystallisation of lovastatin, showing that this more general approach to crystallisation design can be used effectively to optimise desired metrics, exceeding solutions that would otherwise be achieved by decomposition-based methods. Furthermore, it is shown that coupling of cooling and anti-solvent crystallisation can achieve greater crystal yields, whilst reducing the mass of solvent required for the process, further emphasising the usefulness of this approach, and of solvent mixtures in general.

[1] Brown CJ, McGlone T, Yerdelen S, Srirambhatla V, Mabbott F, Gurung R, et al. Enabling precision manufacturing of active pharmaceutical ingredients: workflow for seeded cooling continuous crystallisations. Molecular Systems Design & Engineering. 2018.

[2] Achenie L, Venkatasubramanian V, Gani R, editors. Computer aided molecular design: theory and

practice. Elsevier. Amsterdam, The Netherlands; 2003.

[3] Gani R. Chemical product design: challenges and opportunities. Computers & Chemical Engineering. 2004;28(12):2441-57.

[4] Karunanithi AT, Achenie LE, Gani R. A new decomposition-based computer-aided molecular/mixture design methodology for the design of optimal solvents and solvent mixtures. Industrial & engineering chemistry research. 2005;44(13):4785-97.

[5] Karunanithi AT, Achenie LE, Gani R. A computer-aided molecular design framework for crystallization solvent design. *Chemical Engineering Science*. 2006;61(4):1247-1260

[6] Raman R, Grossmann IE. Modelling and computational techniques for logic based integer programming. Computers & Chemical Engineering. 1994;18(7):563-78.

[7] Jonuzaj S, Akula PT, Kleniati P, Adjiman CS. The formulation of optimal mixtures with generalized disjunctive programming: A solvent design case study. AIChE J. 2016;62(5):1616-1633

[8] Jonuzaj S, Adjiman CS. Designing optimal mixtures using generalized disjunctive programming: Hull relaxations. Chemical Engineering Science. 2017; 159 106-130.

[9] Watson OL, Galindo A, Jackson G, Adjiman CS. Computer-aided design of solvent blends for the cooling and anti-solvent crystallisation of ibuprofen. European Society of Computer-Aided Process Engineering. 2019, submitted

[10] Hutacharoen P, Dufal S, Papaioannou V, Shanker RM, Adjiman CS, Jackson G, Galindo A. Predicting the solvation of organic compounds in aqueous environments: from alkanes and alcohols to pharmaceuticals. Industrial & Engineering Chemistry Research. 2017;56(38):10856-76.

568701 Two-Dimensional Moisture Content and Size Evolution during Fluid Bed Drying As Part of a Continuous Pharmaceutical Manufacturing Process Using NIR-CI

The pharmaceutical industry is currently making the transition from batch to continuous manufacturing for the latter method's benefits of e.g. process flexibility, less waste of materials and improved control of the product. For the manufacturing of oral solid dosages, the innovative ConsiGma-25™ continuous wet granulation line applies twin-screw granulation for production of granules for qualitative tableting. To enable downstream processing, the line uses a semi-continuous fluid bed dryer which reduces the excess liquid in the wet granules coming from the twin-screw wet granulation unit. This process is robust, yet features a multitude of factors that make the dryer system complex. Firstly, the dryer distributes a continuous stream of wet granules over its six drying cells. The cells are sequentially filled for a fixed amount of time, after which the granules are dried and discharged. Due to the gradual filling of each cell, the wet granules start drying at different points in time and the drying regime changes over time. Furthermore, critical process parameters such as the inlet air flow rate, inlet air temperature and cell filling time influence the fluidization and evaporation behavior of the material in the dryer cell. Finally, also the granule size has a major influence on the drying behavior (De Leersnyder et al., 2018).

The study of this complex process, with the aim of Design Space determination in the Quality-by-Design paradigm of the pharmaceutical manufacturing industry (ICH, 2009), is therefore best performed by means of mechanistic modelling. In this way the various interacting phenomena such as fluidization behavior, single particle-level drying and conditions of the drying air can be captured in order to predict the drying behavior of the key product. In models with many interacting factors such as fluid bed drying, sufficiently detailed data is necessary to calibrate and validate the model, to ensure that the final model is the correct representation of the process. As such, these drying models do predict a moisture content distribution in the material (Ghijs et al., 2019; Peglow et al., 2007). These arise from fluidization behavior, the continuous filling of a dryer cell and the heterogeneity in granule size encountered in pharmaceutical granules. Yet in practice, the only available product data is the average moisture content of the granules, meaning that the model cannot compare its output to reality in full extent. This makes that some parts of the model might be insufficiently validated with the reality, and therefore are not able to extrapolate to other process conditions or systems. With two-dimensional moisture content and size distribution data, these models instead would be able to compare their output one-to-one to experimental data, and more components of the model (effect of fluidization, granule size, air conditions) can be validated, which enhances its predictive power.

Therefore, in this study an at-line analytical method was developed to simultaneously determine the moisture content as well as size of individual pharmaceutical granules to construct a 2D distribution. Spatially distributed moisture content measurements of the granules were obtained by dispersing the granules on a conveyor belt (ENP, Hjälteby, Sweden) with a Retsch DR 100 vibrating feeder (Retsch, Haan, Germany), which passed the granules underneath a hyperspectral camera: a near-infrared chemical imaging (NIR-CI) device (VLNIR with OLES56 lens, Specim, Finland). Method validation experiments were performed by measuring multiple samples of granules first with the NIR-CI set-up, immediately followed by a reference LOD instrument (Mettler LP16, Mettler-Toledo, Zaventem, Belgium). By drying wet granules (i) in open air, (ii) in a 50°C oven and (iii) in a fluid bed dryer (Glatt GPCG1), granules with different levels of moisture content were targeted for method validation. All hyperspectral data was then processed in MATLAB (MathWorks®, Natick, Massachusets, USA) using the HYPER-tools package (Mobaraki & Amigo, 2018). The obtained spectra were processed in order to analyze the granules captured on the image. To relate the granules' spectra to the actual moisture content, the average spectrum of the granules on each image was obtained and compared to the LOD measurement of the sample. A partial least squares (PLS) model was used to derive the moisture content from the spectra. Good model predictions were obtained), and thus the spectra could accurately indicate the actual moisture content, and this even for granules larger than 3 mm.

In the same setup, the size of each granule on the image is determined, and by means of the PLS-model the moisture content of each granule is found, resulting in a two-dimensional moisture content and size distribution. These distributions capture the spatial distribution of the moisture in the key product. By analyzing the distribution at different drying times and process conditions, highly accurate information is obtained on the evolution of moisture content of the material. In conjunction with a 2D population balance model, this will allow to gain more in-depth knowledge of the drying process. The latter is planned as next step and is out of scope of this abstract which focuses on the data collection.

References

De Leersnyder, F., Vanhoorne, V., Bekaert, H., Vercruysse, J., Ghijs, M., Bostijn, N., ... De Beer, T. (2018). Breakage and drying behaviour of granules in a continuous fluid bed dryer: Influence of process parameters and wet granule transfer. *European Journal of Pharmaceutical Sciences*, *115*, 223–232. http://doi.org/10.1016/j.ejps.2018.01.037

Ghijs, M., Schäfer, E., Kumar, A., Cappuyns, P., Van Assche, I., De Leersnyder, F., ... Nopens, I. (2019). Modeling of Semicontinuous Fluid Bed Drying of Pharmaceutical Granules With Respect to Granule Size. *Journal of Pharmaceutical Sciences*, 1–8. http://doi.org/10.1016/j.xphs.2019.01.013

ICH. (2009). *International Conference on Harmonisation (ICH) of Technical Requirement for Registration of Pharmaceuticals for Human Use, Pharmaceutical Development, Q8 (R2).*

Mobaraki, N., & Amigo, J. M. (2018). HYPER-Tools. A graphical user-friendly interface for hyperspectral image analysis. *Chemometrics and Intelligent Laboratory Systems*, *172*(October 2017), 174–187. http://doi.org/10.1016/j.chemolab.2017.11.003

Peglow, M., Kumar, J., Heinrich, S., Warnecke, G., Tsotsas, E., Mörl, L., & Wolf, B. (2007). A generic population balance model for simultaneous agglomeration and drying in fluidized beds. *Chemical Engineering Science*, *62*(1–2), 513–532. http://doi.org/10.1016/j.ces.2006.09.042

571347 Continuous Mixing Technology: Design Space Exploration with Discrete Element Simulations

Continuous Mixing Technology: Design Space Exploration with Discrete Element Simulations

Peter Toson[1], Pankaj Doshi[2], Eva Siegmann[1], Johannes Khinast[1,3], Daniel Blackwood[2], Ashwinkumar Jain[2], Alexandre Bonnassieux[2], Dalibor Jajcevic[1]

[1] Research Center Pharmaceutical Engineering, Inffeldgasse 13, 8010 Graz, Austria

[2] Worldwide Research and Development, Pfizer Inc. Groton CT 06340

[3] Graz University of Technology, Institute of Process and Particle Engineering, Inffeldgasse 13, 8010 Graz, Austria

Keywords: Pharmaceutical Manufacturing; Continuous Mixing Technology; Process Design & Development

Continuous manufacturing has many advantages over batch processing, including smaller footprints, faster turnaround times, and better control of process parameters and product quality. A minimal setup for a continuous tableting line consists only of three unit operations (direct compression): feeding of the individual components, mixing and tableting. Many critical quality attributes (CQAs) of the final tablet depend on the mixing step: the mixing quality directly influences the content uniformity of the tablet, the extent of lubrication influences the tensile strength of the final tablet.

The aim of this work is an in-depth analysis of a vertical continuous mixing device termed Continuous Mixing Technology (CMT, see Fig. 1). The vertical design allows to control the impeller speed (and thus the shear rates) independently from the hold-up mass (and thus mean residence time). The same number of blade passes can be generated with different impeller speed and hold-up mass combinations. Thus, the vertical design of the CMT opens up more degrees of freedoms when it comes to process design compared to the traditional horizontal mixing devices, where the screw speed directly influences both shear rate and mean residence time.

Discrete Element Method (DEM) simulations have been performed using the commercial software package XPS (Extended Particle System). The simulations model production-scale processes with a hold-up mass of up to 1000g and mass throughputs ranging from 30kg/h to 40kg/h with 5.7million particles. The model has been validated with tracer experiments in a previous study [1].

A wide range operating conditions have been examined with DEM simulations. Simulation results elucidate the powder mixing dynamics as captured by residence time distribution throughout the operating space. It helps in identifying most optimal operating space. Thus virtual design space exploration with DEM simulations has been proving to be a valuable tool in process development.

References:

[1] Toson et al., Int. J. Pharm. 552, 1–2 (2018), 288–300. doi:10.1016/j.ijpharm.2018.09.032.

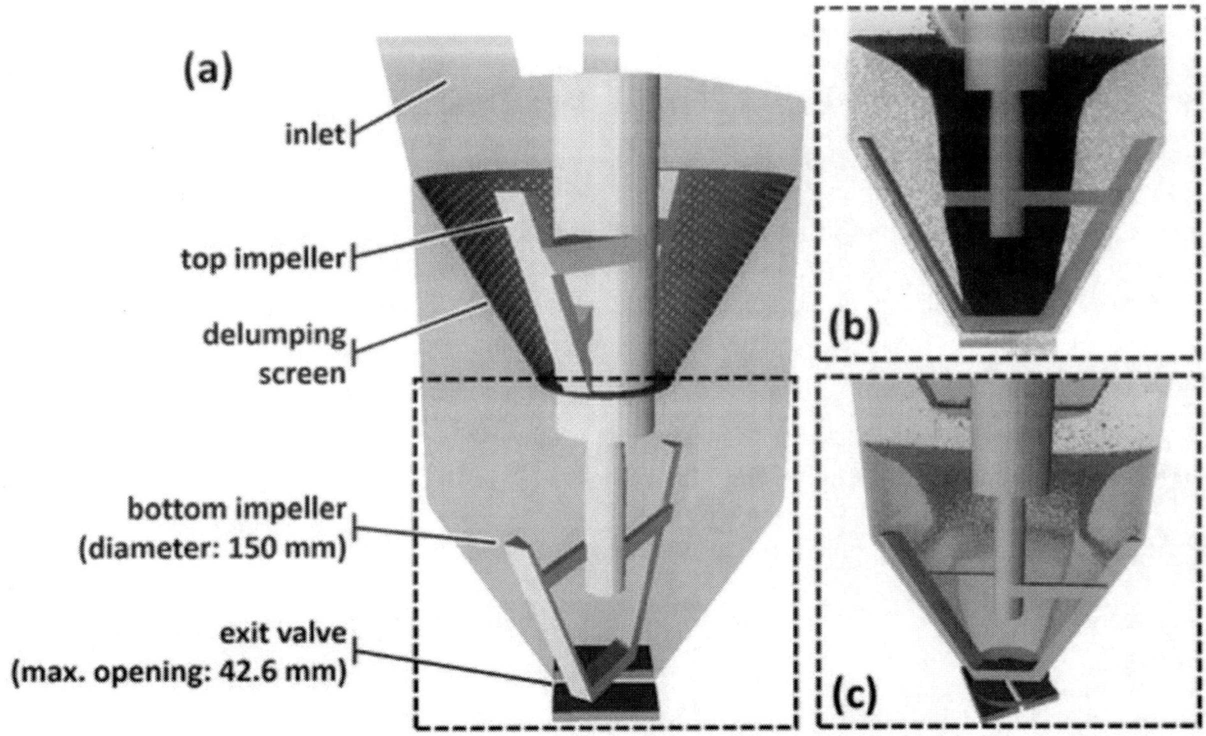

Figure 1. (a) CMT overview. (b) A cut through the CMT to reveal the powder bed shape in the mixing zone at 350 rpm. (c) Tracer particles (red) used to determine the residence time distribution. Snapshot taken 2s after insertion of the tracer impulse.

573794 Validation of Combined PBM-RTD Technique to Simulate Continuous Wet Granulation Systems for Pharmaceutical Manufacturing Processes

Wet granulation is a key unit operation in the pharmaceutical industry employed in the production of solid oral dosage formulations through the continuous manufacturing route. In an effort to predict the outlet compositions of the particles of a continuous granulator, the study of residence time distributions (RTDs) is crucial in order to predict the effects of variabilities in the inlet parameters such as feeder fluctuations and blend inhomogeneity on the outlet granule particle sizes and compositions. Population Balance Models (PBMs) are developed and extensively employed in modeling wet granulation systems in order to track the particle size distributions (PSDs) and porosities of the particles [1].

An initial experimental study on a twin screw granulator was performed using a low API dose blend across a DOE consisting of variation in powder feed rate, liquid content and screw rotation speed. The RTD pulse experiments were carried out on the DOE and exit concentration was measured in real time using an inline near infrared probe (NIR). The residence time profile was fitted to a theoretical model consisting of tank-in-series, plug flow fractions, parallel streams, etc. Moreover, empirical mathematical correlations were obtained for the mean residence time (MRT) with respect to the operating DOE conditions. These studies and results were in line with previously published results in literature [2, 3].

A granulation PBM with RTD mechanics incorporated to it would track the flow of key species such as API thereby quantifying the processes and attributes of the outlet granules completely. The model would provide the researcher with the potency of the API across granule sizes. The goal of this proposed study would be to experimentally validate the developed models. The results of the study would help in developing real-time control strategies that would ensure granules with critical quality attributes (CQAs) consistent over time during the production of the granules.

[1] A. Chaudhury, 'Mechanistic modeling, simulation and optimization of wet granulation processes', Rutgers University - Graduate School - New Brunswick, 2015.

[2] A. Kumar et al., 'Conceptual framework for model-based analysis of residence time distribution in twin-screw granulation', European Journal of Pharmaceutical Sciences, vol. 71, pp. 25–34, Apr. 2015.

[3] H. Y. Ismail, M. Singh, S. Darwish, M. Kuhs, S. Shirazian, D. M. Croker, M. Khraisheh, A. B. Albadarin, and G. M. Walker, 'Developing ann-kriging hybrid model based on process parameters for prediction of mean residence time distribution in twin-screw wet granulation,' Powder Technology, vol. 343, pp. 568-577, 2019.

566521 Computational Modeling of Liposome Formation in a Continuous Jet Flow Process Using CFD and MD

Purpose: Liposomes are spherical vesicles composed of a bilayer and/or a concentric series of multiple bilayers separated by aqueous compartments formed by amphipathic molecules such as phospholipids that are commonly used as carriers for drug delivery purposes. The current study focused on the continuous liposome processing system developed at UConn, where two liquid flows are mixed under a controlled manner through a coaxial jet in co-flow arrangement, of which the center flow consists of an ethanol solution of lipids and the exterior flow is an aqueous solution. It was suspected that the intermolecular forces among various components (e.g., ethanol-lipid-water) along with the unique fluid dynamics (e.g., jet flow) caused lipid aggregation and subsequent liposome formation. However, the underlying mechanism as well as the process dynamics are yet fully understood. Accordingly, we have carried out a multi-scale computational study of liposome formation in coaxial turbulent jet flow to probe the underlying mechanism and to quantitatively predict the formation of liposomes.

Methods: Both computational fluid dynamics (CFD), as a macro-scale simulation, and coarse-grained molecular dynamics (CG-MD), as a micro-scale investigation, have been conducted to not only reveal the detailed mechanism of liposome formation, but also implement multi-scale case studies for the process. CFD simulations were verified by comparison of flow pattern as well as formation temperature with experiments.

Results: The CG-MD simulations revealed that MARTINI force field (FF) could not capture a realistic behavior of lipids and cholesterol in ethanol solution due to lipids aggregation, which caused deviation from both the experimental results and all-atom MD simulations. Further optimization of MARTINI FF with reference to all-atom MD simulations was performed using a versatile object-oriented toolkit for coarse-graining applications (VOTCA). The multi-scale simulations were found to be most accurate when compared to the experimental data and trends. The CFD simulation results suggested that including heat of mixing in the energy equation would be desired to obtain a formation temperature comparable with the experiments.

Conclusions: Default MARTINI potential energy parameters require optimization with reference to an all-atom MD simulation for properly modeling lipids in ethanol solution. Discrete and continuum computational modeling of liposome formation were found to be two viable complementary approaches which cover micro-scale and macro-scale, respectively.

Acknowledgements: FDA Grant# 1U01FD005773-01.

Disclaimer: This article reflects the views of the authors and should not be construed to represent FDA's views or policies.

579336 A Big Data Analytics Workflow for Pharmaceutical Manufacturing Industry

Pharmaceutical manufacturing is a complex process, consisting of blending, granulation, tableting and other unit operations. The aim of pharmaceutical manufacturing is to reliably produce good quality drugs with high efficiency and low cost. However, due to lack of process understanding and control strategies, it is hard to avoid drug quality defects and production efficiency decreasing. Nowadays, many pharmaceutical manufacturing unit operations have been equipped with lots of sensors, controllers and even process analytical instruments, and this has enabled the rapid growth of pharmaceutical manufacturing industry big data. The vast data resource hides the process information and characteristics, so mining knowledge behind the manufacturing data is of vital importance.

In this study, we proposed a big data analytics workflow for pharmaceutical industry manufacturing to mine process knowledge, which helped to improve drug quality and production efficiency. The workflow contains five steps: problem definition, data acquisition, data preprocessing, data modelling and model application. The workflow provides an idea for pharmaceutical enterprises to collect, organize and analyze manufacturing data systematically. Meanwhile, a data analytics case study of a double effect evaporation process for herbal medicines was used for further illustration. In the case, the production efficiency of a double effect evaporator was observed to decrease during a long time. To handle this problem, the data of sensors, valves and instructions from 172 batches were collected. After data preprocessing, data modelling task was carried out. Correlation analysis was first applied to study the relationships among each single process variables. Next, multivariate data analysis methods were used to evaluate the profiles of all the batches: different process phases were identified using hierarchical clustering analysis according to the dissimilarities among the batches; the root causes of the phase transformations were also investigated through principal component analysis. The models above could be applied for on-line process monitoring, thus helping to enhance process understanding and to support manufacturing decision.

574446 An Efficient Data-Based Methodology to Identify the Design Space of Continuous Pharmaceutical Manufacturing Processes

Flowsheet modeling[1,2] is an increasingly accepted tool to simulate and improve continuous processes for solid oral dosage manufacturing. Advances in continuous manufacturing have led to increased focus on developing accurate and detailed models for the unit operations. Flowsheet models developed using the unit models to simulate dynamic behavior of the continuous process are computationally expensive. Strategies are needed to overcome this limitation and fully realize the potential of these models. Specifically, the models can be utilized to enable decision making during detection of off-spec product, risk assessment and process optimization.

Tools such as sensitivity analysis are established[3], and can be applied to flowsheet models to help identify the critical process parameters. While this helps in identifying variables where further research efforts can be focused on, the resulting problem is still high dimensional for implementing advanced analyses such as identification of design space. A design space is characterized by the range of variables within which the process satisfies equipment, quality and production constraints. Traditional methods to identify the design space require high sampling cost or closed form constraints, which limits its use with flowsheet models. Hence, effective strategies are required to handle problems typically encountered in pharmaceutical processes.

Current work uses surrogate based feasibility analysis method[4] that builds surrogate models to approximate the feasibility function that characterizes the maximum constraint violation. This strategy uses a modified expected improvement function to identify samples close to the feasible region boundaries and unexplored regions. Previous work published used kriging[4] and radial basis function[5] as the surrogate models, which suffered limitations for high dimensional problems. Specifically for flowsheet models, the design space was identified using a set of two dimensional problems, which ignores interactions between the variables. In this work, a novel artificial neural network (ANN) based methodology is used to identify the design space. An ANN based surrogate model is built to approximate the feasibility function through carefully identifying samples using an adaptive sampling strategy. The unexplored regions are identified using the modified expected improvement function and a variance estimator. The variance of an unexplored sample is estimated using a statistical technique known as Jackknifing[6,7]. The developed approach is utilized to effectively identify the design space for an integrated direct compaction as well as wet granulation line that simulates a plant scale operation.

References

1. Boukouvala F, Niotis V, Ramachandran R, Muzzio FJ, Ierapetritou MG. An integrated approach for dynamic flowsheet modeling and sensitivity analysis of a continuous tablet manufacturing process. *Computers & Chemical Engineering*. 2012;42:30-47.
2. Yoon S, Galbraith S, Cha B, Liu H. Flowsheet modeling of a continuous direct compression process. 2018:121-139.
3. Saltelli A, Annoni P, Azzini I, Campolongo F, Ratto M, Tarantola S. Variance based sensitivity analysis of model output. Design and estimator for the total sensitivity index. *Computer Physics Communications*. 2010;181(2):259-270.
4. Boukouvala F, Ierapetritou MG. Derivative-free optimization for expensive constrained problems using a novel expected improvement objective function. *Aiche J*. 2014;60(7):2462-2474.
5. Wang Z, Ierapetritou M. A novel feasibility analysis method for black-box processes using a radial basis function adaptive sampling approach. *Aiche J*. 2017;63(2):532-550.
6. Eason J, Cremaschi S. Adaptive sequential sampling for surrogate model generation with artificial neural networks. *Computers & Chemical Engineering*. 2014;68:220-232.
7. Jin Y, Li J, Du W, Qian F. Adaptive Sampling for Surrogate Modelling with Artificial Neural

Network and its Application in an Industrial Cracking Furnace. *The Canadian Journal of Chemical Engineering*. 2016;94(2):262-272.

565756 Condition Based Maintenance for Sensor Network Reliability in Continuous Oral Solid Dose Manufacturing

The advent of Industry 4.0 (Smart Manufacturing) in the 2010s, is a manifestation of cultural and mindset change across multiple industries towards the acceptance of digital tools in process design and operations, resulting in highly automated and data-rich systems to realize lean manufacturing. These process intelligence tools leverage cross-industry advances in manufacturing methods, data analytics, fault management, and implementation technologies, for sustainable process operations, safety, quality risk assessment, and enhanced customer relations. In pharmaceutical systems, such digitalization tools facilitate the implementation of continuous manufacturing and real-time release testing by tracking product quality in real-time, thereby enabling the release of life critical medication with reduced off-line quality testing compared to existing practices.

Continuous manufacturing of drug product is the outcome of targeted process intensification with systematic integration of product and process knowledge, instrumentation and automation systems, quality control protocols and real-time process management for a solids processing system handling pharmaceutical materials. Success in the technology necessitates appropriate infrastructure for Information and Operations Technology integration.

At the core of such integrated systems is sensor network analytics, consisting of reliable and accurate measurements from field devices combined with mature process understanding to provide good quality estimates of unmeasured process variables [1]. In continuous tablet manufacturing, four leading causes of instrumentation faults are observed, which affect monitoring robustness. These include communication failure, instrument configuration, calibration and instrument age. Such instrumentation failures could potentially cause a systemic failure of the CM process.

This work focuses on leveraging systems engineering methods for real-time operations management, in particular, condition-based maintenance to improve operational reliability by mitigating sensor network failures in drug product CM. We discuss the following: (i) an integrated data-driven and model-based framework for systematic sensor network fault monitoring, (ii) the configuration and implementation of condition-based maintenance in a tablet processing pilot plant at Purdue University [2].

The proposed maintenance management approach is comprised of three workflows which are linked in an integrated data-driven and model-based framework. First, sensor network analytics methods of statistical process monitoring, data reconciliation, and gross error detection are leveraged for real-time process data validation, ensuring measurement accuracy, detecting sensor faults, and estimating unmeasured variables. Proof of concept of these approaches have recently been demonstrated by the research group [3,4]. Second, gross errors are identified for their root cause for appropriate maintenance activities, such as corrective, preventive and predictive maintenance. Third, to reconnect a maintained sensor, guidelines such as ASTM D6122, E2968 are leveraged and customized for the process.

The integrated framework is implemented at Purdue University in a pilot scale continuous direct compression facility. An analytically redundant sensor network using process equipment sensors, inline and at-line sensors are networked with automation systems such as Emerson DeltaV, OSIsoft PI System and Applied Materials Smart Factory Rx. A hierarchical systems integration architecture closely following ISA-95 is set up to advance the practice of Quality-by-Design in continuous manufacturing towards Quality-by-Control and Industry 4.0 [4].

References:

[1] Bagajewicz MJ. Smart Process Plants: Software and Hardware Solutions for Accurate Data and Profitable Operations. McGraw-Hill, 2009

[2] S. Ganesh, G. Reklaitis et al. Leveraging the digital infrastructure for real-time operations management in continuous pharmaceutical manufacturing: Condition-based maintenance for sensor network robustness

(in preparation).

[3] M. Moreno, S. Ganesh, Y. Shah, Q. Su, M. Gonzalez, Z. Nagy, G. Reklaitis. Sensor Network Robustness using Model-based Data Reconciliation for Continuous Tablet Manufacturing. J Pharm Sci (2019).

[4] Q. Su, S. Ganesh, M. Moreno, Y. Bommireddy, M. Gonzalez, G. Reklaitis, Z. Nagy. A perspective on Quality-by-Control (QbC) in pharmaceutical continuous manufacturing. Comput Chem Eng (2019).

569352 Toll Manufacturing: A Sharing-Economy Model for on-Demand Chemical Production Services

A toll manufacturing-based economy is touted as the next era of chemical processing wherein toll manufacturers provide critical on-demand chemical production services in return of a fee or toll. Compared to traditional chemical industries where plant owners are both investors and operators, toll manufacturing is based on a sharing-economy model involving two parties – toll manufacturers and consumers. The services provided by toll manufacturers include specialized equipment, multi-purpose processes and manufacturing facilities whereas consumers provide the necessary raw material and/or intermediate feedstocks. For the consumers, this model is especially advantageous as they can significantly reduce time-to-market, promptly respond to market dynamics and test new feedstocks and products without long-term investments on equipment, labor and expertise.

From toll manufacturers' viewpoint, multi-purpose or flexible processes are preferred during chemical manufacturing as different products can be produced from same or different feedstocks. Typically, multi-purpose chemical plants have been shown to be more economically profitable for specialty chemicals such as pharmaceuticals, paper and polymers. These specialty chemicals are manufactured in small amounts, and therefore, multi-purpose plants have shown promise to react faster to dynamic market conditions. On the other hand, dedicated single-functional plants are more economically attractive for large-scale production of chemicals. Continuing with this understanding, with increasing amounts of small-scale and stranded natural gas resources, the need of multi-functional plants for small-scale gas processing will be apparent in the near future.

In the chemical space, toll manufacturing, in particular, requires process agility and flexibility which can lead to quick process transitions in response to market needs. This feature demands the design and operation of reconfigurable, switchable and flexible plants, which poses significant new challenges in the areas of process design, synthesis, scheduling, and transition control. This presentation will discuss and demonstrate how process systems engineering principles can be applied for addressing some of these challenges.

574556 Integrated Process Development and Economic Evaluations for an Optimum Gas Processing Plant

Heavy hydrocarbon removal is one of the processes in natural gas processing plants for reducing heating value and to control hydrocarbon dew point of sales gas. Conventional technologies include Joule Thomson expansion, turboexpansion, mechanical refrigeration, membrane separation and supersonic centrifugal separation. The aim of this work is to develop and optimize an integrated process for natural gas plant and natural gas liquid (NGL) productions. The system is based on acid gas removal unit, simultaneous absorption of heavy hydrocarbons and water from natural gas using a mixture of triethylene glycol (TEG), NGL recovery and sales gas compression. Natural gas processes are modelled and simulated within Aspen HYSYS simulator with a common set of operating criteria and thermodynamic parameters for the absorption process, based on field data. The reboiler temperature of TEG regenerator and then the hot stripping gas flowrate, have been studied for their response to changes in the regenerated TEG concentration, dew point of sale gases, TEG losses (make-up) and reboiler duty. Finally, Aspen process economic analyzer program is used to calculate the expected capital expenditures of the plant. Process description, operating conditions, equipment sizing, recovery efficiency and limitations for the new absorption process are presented. This integrated technology has an equipment sizing, investment and operating cost, and energy consumption lower than commercial technologies available nowadays. Results indicate an increase in hydrocarbon absorption efficiency. Energy consumption is lower than other processes, while the capital cost and operating cost are also lower.

566980 Modeling of Skinning Effect and Optimization of Dry-Casting Process

Dry casting process is one of the methods to produce a polymer membrane film by casting polymer solution on the continuously moving support film and supplying heated airflow to vaporize solvent form the solution. The polymer membrane film can be produced with merely overheated airflow, but there is probability of occurring skinning effect which deforms the morphology of the membrane film. The main cause of skinning effect is a low solvent concentration near the surface of the polymer solution. When solvent is vaporized with overheated airflow, solvent concentration near the surface is reduced more rapidly than that of the solution due to the sharp vaporization at the surface. If solvent concentration decreases, the diffusion coefficient of solvent in the solution also decreases, which limits the diffusion of the solvent near the surface. As a result, skinning effect occurs; thin solid layer is formed on the top of the solution and hinders solvent inside the solution from evaporating [5, 6]. Therefore, prediction and prevention of skinning effect is an important issue in operating dry-casting process.

The process conditions causing skinning effect can be analyzed through experiments and established empirically. However, it is costly and time consuming to collect sufficient data for analysis. The analysis of skinning effect will be efficient if causal conditions of skinning effect are predictable via model simulation. Therefore, it is aimed to find the process conditions that produce the polymer film that meets the desired value while minimizing the skinning effect. The process in the simulation includes 6 zones through which the polymer solution passes in order and temperatures of these zones are control variables to adjust properties of the final polymer film. The nonlinear model which consists of Flory-Huggins theory and Free volume theory is used for the simulation [1-4]. Flory-Huggins theory accounts for mixing effect of polymer solution different from that of general solutions. Free volume theory describes mass transfer equations of polymer solution and the interactions between solvent and polymer [7, 8]. The system for the simulation is polystyrene/toluene as polymer/solvent. Diffusion coefficient of solvent is traced so as to analyze skinning effect via the simulation since skinning effect cannot be observed directly in the model simulation. First, temperatures of 6 zones are randomly selected to figure out the tendency of diffusion coefficient of solvent to temperature distribution. Trajectories of diffusion coefficient of solvent are described through thickness direction of the polymer solution and temperature distribution where diffusion coefficient of solvent is drastically declined is grasped the trajectories. Furthermore, optimization is also performed based on the model equations to find the temperature distribution that minimizes the curvature of the trajectories and to be compared with the results of random simulations.

Temperature distribution that makes diffusion coefficient of solvent decrease slowly was found through temperature random sampling simulations. Solvent is removed by evaporation on the top of the solution and by diffusion at the bottom. Thus, concentration gradient need to be made by heated airflow so as to diffuse the solvent of the bottom upward. However, overheated airflow accelerates evaporation of the solvent on the surface, decreases diffusion coefficient rapidly and eventually induces skinning effect. Therefore, time for diffusion needs to be given with lowering temperature after concentration gradient is formed to prevent skinning effect. Temperature distribution via optimization also has a similar result, which is high temperature at early stage and relatively low temperature in a next zone. The strategy for prediction of skinning effect could be obtained by model simulation and optimization. Skinning effect in other dry-casting process with different properties of polymer solution is predictable by using the model simulation and the number of experiments to find out temperature conditions for preventing skinning effect is minimized.

References

[1] Altinkaya, S. A., & Ozbas, B. (2004). Modeling of asymmetric membrane formation by dry-casting method. *Journal of Membrane Science*, *230*(1-2), 71-89.

[2] Arya, R. K. (2012). Drying Induced Phase Separation. *Journal of Chemical Engineering*, *27*, 12-20.

[3] Bouyer, D., Werapun, W., Pochat-Bohatier, C., & Deratani, A. (2010). Morphological properties of membranes fabricated by VIPS process using PEI/NMP/water system: SEM analysis and mass transfer modelling. *Journal of Membrane Science, 349*(1-2), 97-112.

[4] Bouyer, D., & Pochat-Bohatier, C. (2013). Validation of mass-transfer model for VIPS process using in situ measurements performed by near-infrared spectroscopy. *AIChE Journal, 59*(2), 671-686.

[5] Ciampi, E., & McDonald, P. J. (2003). Skin formation and water distribution in semicrystalline polymer layers cast from solution: a magnetic resonance imaging study. *Macromolecules, 36*(22), 8398-8405.

[6] Vinjamur, M., & Cairncross, R. A. (2001). A high airflow drying experimental set-up to study drying behavior of polymer solvent coatings. *Drying Technology, 19*(8), 1591-1612.

[7] Vrentas, J. S., & Vrentas, C. M. (1995). Determination of free-volume parameters for solvent self-diffusion in polymer-solvent systems. *Macromolecules, 28*(13), 4740-4741.

[8] Vrentas, J. S., & Vrentas, C. M. (2003). Evaluation of the free-volume theory of diffusion. *Journal of Polymer Science Part B: Polymer Physics, 41*(5), 501-507.

553828 Process Considerations to Maximize Catalyst Activity in Solution Polyolefin Polymerizations

Production of polyolefins by coordination catalysts in solution reactors is of increasing industrial importance. The development of novel, complex catalyst structures is improving the performance of existing products and expanding the market into new applications. However, as the complexity of these catalyst structures increases so too does the cost. In order to maintain reasonable manufacturing costs, greater attention must be given to maximizing the activity of these catalysts in the polymerization reactor. This paper reviews chemical processing techniques to maximize catalyst activity in solution polyolefin, polymerization reactors.

568754 Carbonate Fuel Cells: Scaling-up the Magic Carbon Capture Box

To economically sequester or utilize CO_2 from dilute sources such as industrial emissions or power generation, technologies for concentrating CO_2 are required. Unfortunately, commercially available technologies are capital and energy intensive, reducing their effectiveness in limiting emissions. The advantage of the carbonate fuel cell is that it can effectively concentrate CO_2 and simultaneously produce electricity and hydrogen. This example of process intensification takes advantage of the electrochemical reaction to drive a desirable separation. Although carbonate fuel cells have been deployed commercially for power generation, strenuous conditions of carbon capture are much more challenging. A joint development program has been created to reengineer the fuel cell stack for carbon capture, utilizing model-guided and parallel development strategies.

Work on single cells has shown that the use of dilute CO_2 feeds from flue gas sources can have a significant effect on molten carbonate fuel cell behavior beyond increased cathode overpotentials. Specifically, testing has shown that the electrolyte can behave as a dual ion conductor passing both carbonate and non-carbonate ions through the electrolyte. Kinetic testing has helped elucidate mechanistic reaction pathways and suggest that the non-carbonate ions are likely OH^- formed from the catalysis of H_2O and O_2. In addition, mass transport effects are found to be of increased importance at these conditions and a leading factor in controlling fuel cell ionic selectivity. These learnings highlight the need for several design modifications which can lead to improved fuel cell performance in carbon capture applications.

A kW scale technology stack with commercially sized repeat components is being used to evaluate the scale-up challenges of mass transfer and thermal management. Insights from the lab scale fundamentals research have been applied to design a cathode current collector that reduced the CO_2 mass transfer limitations in the fuel cell cathode. With a redesigned current collector, the stack ionic selectivity was improved by >15% while simultaneously increasing voltage by >40 mV. Additionally, because of the exothermic nature of the electrochemical reactions and electrical resistivity, thermal management is critical. Endothermic steam methane reforming is used to heat balance the fuel cell stack and provide hydrogen fuel. Carefully co-locating the heat generation and consumption reactions is critical to reducing thermal gradients rather than exacerbating them. A modeling approach was utilized to design a methane reforming catalyst pattern that could reduce thermal gradients by ~40% within the fuel cell stack and allow for improved electrochemical performance.

The carbonate fuel cell carbon capture technology is able to reduce parasitic energy demands of carbon capture while actually co-producing hydrogen and electricity. The deployment of this unique reactive separation technology, however, relies on careful engineering to properly manage the heat and mass transfer during scale-up. Fundamentals based models have been key in guiding the scale up by highlighting the key mass and heat transfer issues. Going forward, the model guided design approach will be key in matching process conditions and hardware design to specific carbon capture applications.

570292 Building Block-Based Work and Heat Exchanger Network Synthesis (WHENS)

Energy conservation has been a long-standing grand challenge in chemical industry. Efforts in heat exchanger network synthesis facilitate saving in energy consumption and total annual cost [1-2]. Recently, work and heat exchanger network synthesis (WHENS) has emerged as a promising research area to capture the interplay of pressure and temperature manipulation operations [3-5]. Current optimization-based approach for WHENS leverages on superstructure representations such as state-space representation [6], multi-stage superstructure [7] and representations involving heuristics [8]. However, there exist several limitations for these superstructure representations addressing WHENS. Firstly, equipment configurations need to be postulated based on existing knowledge of unit operations, engineering experience and heuristics. Secondly, classic superstructure representations fail to discover novel pathways with process intensification. Process intensification refers to significant reduction of equipment sizes, waste generation, and increase of productivity [9,10]. In the context of WHENS, multi-stream heat exchanger is a relevant example that reduce the number of equipment while improving the overall performance of WHENS.

To this end, we provide a new representation approach to capture many process configurations of WHENS with innovative pathways [11]. The new representation approach centers around two fundamental elements of abstract building blocks. Firstly, the block interior is used to represent splitting, mixing, utility cooling and utility heating of individual streams. Secondly, the block boundaries between adjacent blocks allow energy flow in the form of heat and work. An unrestricted boundary does not allow pressure and composition changes between adjacent blocks. A semi-restricted boundary is assigned with expansion/compression operations connected through either common (integrated) or dedicated (utility) shafts. A completely restricted boundary allows heat flow driven by temperature gradient across the boundary. We demonstrate the rich process information embedded in the proposed representation through literature examples. The novelty of this approach is that there is no need to specify the stage number as requested by the classic approach. Besides, the proposed representation approach allows automatic generation of numerous work and heat exchanger network with intensification opportunities. We formulate this building-block based WHENS as a mixed-integer nonlinear optimization (MINLP) problem with the objective as minimizing total annual cost. We use several literature examples on liquified energy chain to demonstrate the capability of the proposed synthesis approach.

References:

[1] Furman, K.C.; Sahinidis, N.V. A critical review and annotated bibliography for heat exchanger network synthesis in the 20th century. *Industrial & Engineering Chemistry Research*. 2002, 41, 2335–2370.

[2] Nair, S. K., & Karimi, I. A. Unified Heat Exchanger Network Synthesis via a Stageless Superstructure. *Industrial & Engineering Chemistry Research*. 2019. DOI: 10.1021/acs.iecr.8b04490.

[3] Yu, H.; Fu, C.; Vikse, M.; Gundersen, T. Work and heat integration—A new field in process synthesis and PSE. *AIChE Journal*. 2018.

[4] Fu, C.; Vikse, M.; Gundersen, T. Work and heat integration: An emerging research area. *Energy*. 2018, 158, 796–806.

[5] Huang, K.; Karimi, I. Work-heat exchanger network synthesis (WHENS). *Energy*. 2016, 113, 1006–1017.

[6] Wechsung, A.; Aspelund, A.; Gundersen, T.; Barton, P.I. Synthesis of heat exchanger networks at subambient conditions with compression and expansion of process streams. *AIChE J.* 2011, 57, 2090–2108.

[7] Nair, S.K.; Nagesh Rao, H.; Karimi, I.A. Framework for work-heat exchange network synthesis (WHENS). *AIChE J.* 2018, 61, 871–876.

[8] Uv, P.M. Optimal Design of Heat Exchanger Networks with Pressure Changes. Master's Thesis, NTNU, Trondheim, Norway, 2016.

[9] Tian, Y.; Demirel, S.E.; Hasan, M.M.F.; Pistikopoulos, E.N. An Overview of Process Systems Engineering Approaches for Process Intensification: State of the Art. *Chem. Eng. Process.-Process Intensif.* 2018, 133, 160–210.

[10] S.E. Demirel; J. Li; M.M.F. Hasan. Systematic Process Intensification. *Current Opinion in Chemical Engineering*, 2019, https://doi.org/10.1016/j.coche.2018.12.001.

[11] J. Li; S.E. Demirel; M.M.F. Hasan. Building Block-Based Synthesis and Intensification of Work-Heat Exchanger Networks (WHENS). 2018. *Processes*, 7(1), 23.

567580 Photocatalytic Synthesis of Benzylic Amines Using Metal-Free Dyes: Investigating the Reaction Performance with a Spinning Disc Reactor and a Microchannel Reactor

-

Solar photocatalysis has grown as a green alternative for the synthesis of fine organic compounds. Recently, metal-free dyes have gained interest as visible light photocatalysts as they are more environmentally benign than transition metals, they are cheap and they have long excited-state lifetimes.[1] Eosin Y (characteristic peak 539 nm) is an anionic dye, which has efficiently catalysed a range of reactions including cyclisation, halogenation and redox reactions.[2,3] Despite the advances in the lab-scale synthesis, there is a lack of knowledge on how to scale up these visible light photocatalytic reactions for industrial implementation. To absorb light efficiently, thin films and large surface areas are needed. To date, most photocatalytic reactors have been designed based on enclosed UV-light and are therefore not suitable for visible light. Neither are traditional reactors, such as batch or packed reactors, and thus new process intensification (PI) reactors need to be designed, optimised and evaluated to maximise the light efficiency and to overcome the mass transfer limitations.

The aim of this study is to evaluate two different types of thin film/large surface area reactors (Fig 1) for the intensification of visible light photocatalysis: (i) Spinning Disc Reactor (SDR) and (ii) Microchannel reactor (MCR). A specially designed solar light simulator (40cm*40cm) was used to carry out the oxidative homocoupling of benzylamine, chosen as a model reaction.

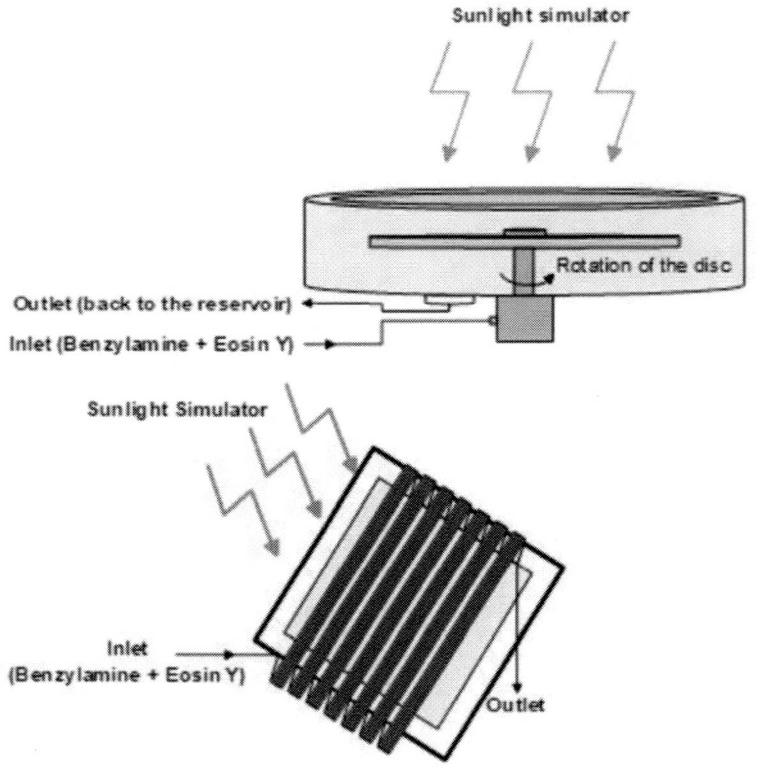

a) b)

Figure 1: Schematic representation of a) Spinning Disc Reactor (SDR); b) Microchannel reactor (MCR).

A range of conditions with low and high substrate concentrations and catalyst loadings were tested with both configurations. As opposed to batch reactors, the photon-transfer limitation was overcome due to the creation of thin films by the centrifugal acceleration in the SDR, and the large surface area per volume ratio in the MCR. Moreover, thin films and high surface areas reduce the effect of the opacity of the solutions, as the light is able to penetrate into the mixture. Therefore, high catalyst loadings and amine concentrations can be used to improve the productivity of the reaction. In Fig 2a, it can be observed that for the MCR, at a high amine concentration, higher catalyst loadings increased the conversion. The productivity also increased as, for example, with 6 mol% Eosin Y and 5 mM initial concentration of amine, 0.41 mmol of amine reacted to form the product whereas with 15 mM, 1.21 mmol of amine reacted.

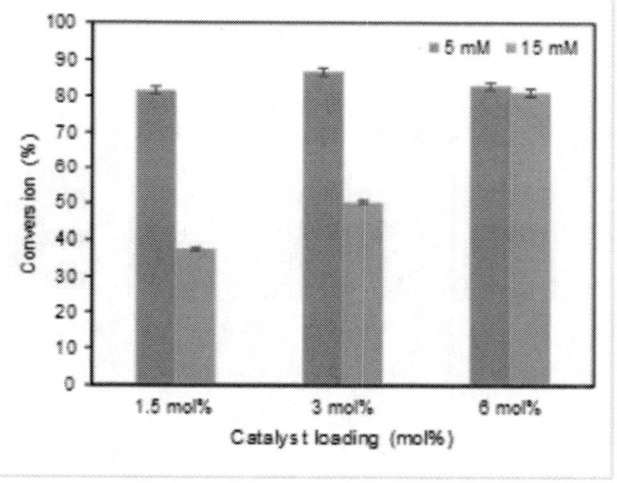

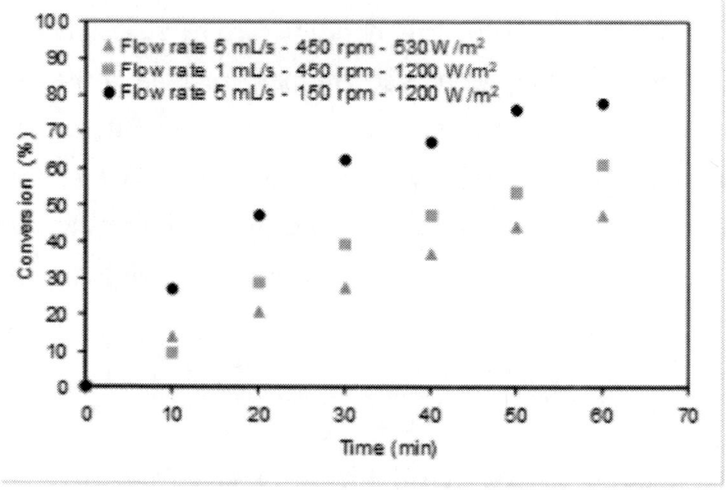

a) b)

Figure 2: a) Effect of catalyst loading and amine concentration using the MCR; b) Effect of flow rate and spinning speed using the SDR.

One of the most significant factors in the performance of the SDR, similarly to the MCR, is the interaction between the flow rate and the light intensity (Fig 2b). This will determine the time that the catalyst spends under the light, in the catalyst-excited state. High flow rates permit a faster recirculation of the solution and therefore, the reaction mixture is on top of the disc more times during a specific reaction time. Due to the thin films created by the SDR (film thickness of 0.09 – 0.45 mm), the light is able to penetrate into the solution and the spinning speed does not greatly influence the conversion.

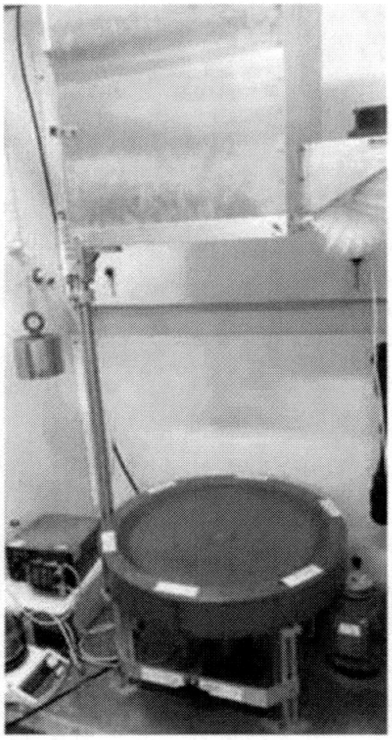

Figure 3. Setup of the solar light simulator and the SDR.

These studies have demonstrated that the MCR and the SDR have the potential to enable intensification of reactions with commercial importance. Further investigations are being carried out with these reactor configurations and with a batch reactor. Moreover, the performance of the SDR (Fig 3) is being investigated with an immobilised catalyst, which creates a process where the catalyst can be simply recovered and reused. In summary, the combination of metal-free dyes and novel visible light photocatalytic reactors have shown to be a solution to maximise the light efficiency, overcome the mass transfer limitations and create a more sustainable process to synthesise fine organic chemicals.

REFERENCES

1 J. R. Xin, J. T. Guo, D. Vigliaturo, Y. H. He and Z. Guan, *Tetrahedron,* 2017, **73**, 4627-4633.

2 D. Prasad Hari and B. König, *Chem. Commun,* 2014, **50**, 6688-6699.

3 J. D. Tibbetts, D. R. Carbery and E. A. C. Emanuelsson, *ACS Sustain. Chem. Eng.*, 2017, **5**, 9826-9835.

564044 Oxidation of L-Methionine in a Rotor-Stator Spinning Disc Reactor

Oxidation of L-Methionine in a Rotor-Stator Spinning Disc Reactor

Arnab Chaudhuri[1] , Koen Kuijpers[1], Parimala Shivaprasad[1,2], Emma Anna Carolina Emanuelsson[2], Timothy Noël[1], John van der Schaaf[1]

[1]Department of Chemical Engineering, TU Eindhoven, 5612 AZ, Eindhoven, Netherlands

[2]Department of Chemical Engineering, University of Bath, BA2 7AY, Bath, United Kingdom

Process intensification (PI) is a promising pathway to achieve sustainable chemical production while also improving the safety of traditional chemical processes [1]. The rotor-stator spinning disc reactor (rs-SDR) is one such PI reactor which consists of a rotating disc enclosed by a stator. Reaction intensification is caused by the high shear force on the feed as a result of the velocity gradient between the rotor and the stator; resulting in high rates of mass and heat transfer [2]. This leads to rs-SDR being a particularly interesting reactor for photochemical applications where photon transfer and mass transfer of the activated species are often limited (e.g. for high concentration). In this study, we demonstrate the potential of a photo-rs-SDR for applications towards photochemical reactions. The oxidation of L-methionine with methylene blue as the catalyst (Figure.1) was chosen as the model reaction. The gas-liquid mass transfer limitations[3] of this systems makes it particularly applicable for the photo-rs-SDRs and demonstrates the potential for improvements in productivity.

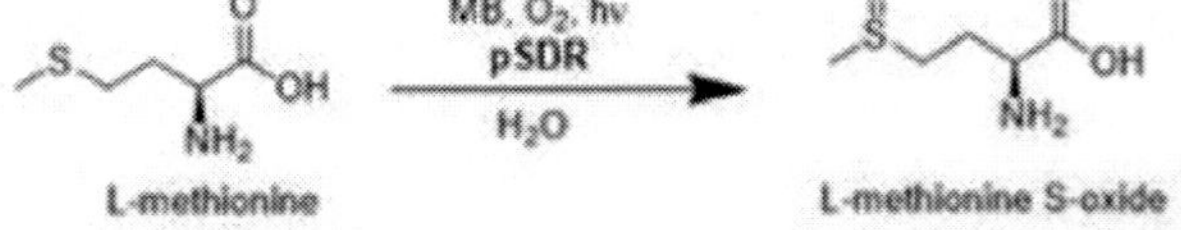

Figure 1: Reaction scheme

The rs-SDR was illuminated using a solar light simulator (AM1.5G), to ensure no light limitations. The reaction was carried out by co-feeding the reaction solution and oxygen at the bottom of the reactor. Samples were collected after a single pass through the reactor and the reaction conversion was measured using HPLC.

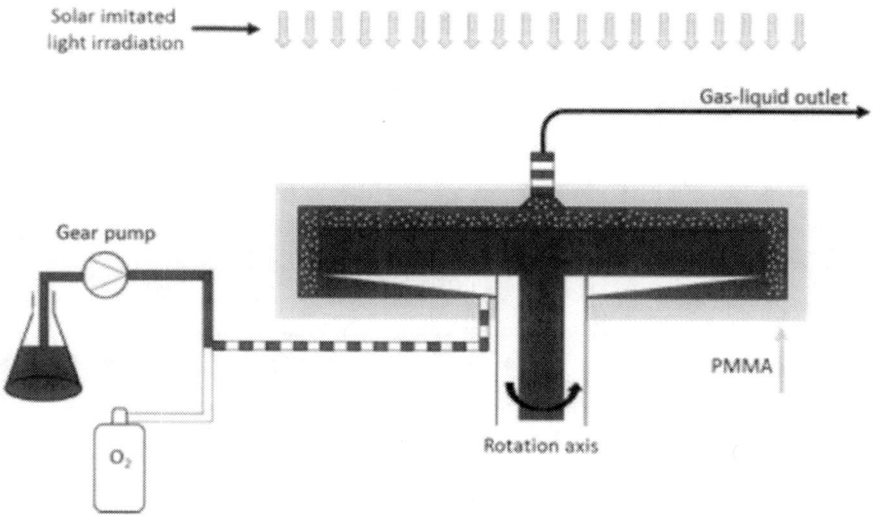

Figure 2: Schematic representation of the reactor set-up

A few of the results obtained are presented in this abstract. Two different configurations were used for the photo-rs-SDR, one with illumination on the dispersed region and one with illumination on the thin film side. As illustrated in Fig 3, the dispersed region seemed to provide better conversions. This is most likely since the mass transfer rates are higher in this region and this allows for more oxygen in the liquid phase, leading to improvements in reaction rates. As the liquid flow rates are increased for constant gas flowrates, the conversions obtained are quite low regardless of the reactor configuration.

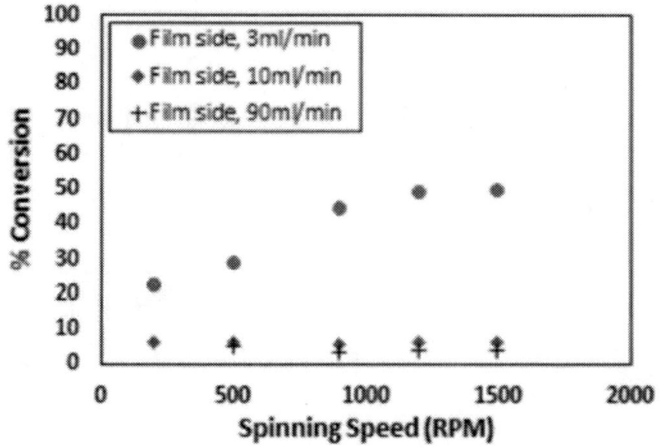

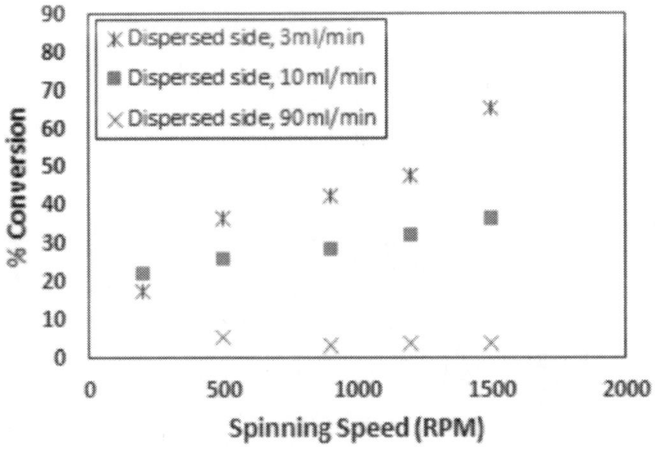

Figure 3: Effect of spinning speed and gas-liquid flowrate on reaction conversion

Figure 4: Visual studies of flow regimes in the photo rs-SDR indicated that uniform distribution of gas could be achieved with increased rotation speed.

Visual studies (Fig 4) of the reactor illustrated that for 1:1 gas to liquid flow ratios, uniform distribution of gas could be achieved throughout the illuminated section of the reactor for the dispersed configuration. This further indicates that the contact between gas and liquid phases in this region was quite high and improved with rotation speeds.

Quite high conversions were achieved in this reactor with residence times of 10 mins. This indicates that the productivity of L-methionine oxidation can be improved. The effect of rotation speed, gas-liquid flowrate and catalyst concentration were investigated on the reaction conversion. We have also investigated these parameters under more light limiting conditions. Additionally, the results were compared to two conventional reactors which are currently used in photochemistry: batch and microflow.

To the best of the authors' knowledge, this is the first study of a successful synergistic reaction system using photochemistry with the rs-SDR. We are currently investigating the potential of the photo rs-SDR for targeted applications in pharmaceuticals and the fine chemical industries.

References

[1] Vlachos, D.G. and S. Caratzoulas, Chemical Engineering Science **2010**, 18-29.

[2] Meeuwse, M., J. van der Schaaf, and J.C. Schouten, *AIChE Journal* **2012**, p. 247-255.

[3]Emmanuel, N., Mendoza, C., Winter, M., Horn, C. R., Vizza, A., Dreesen, L,& Monbaliu, J. C. M., *Organic Process Research & Development*, **2017**,p. 1435-1438.

569946 Development of Sorption Enhanced Reactive Processes for the Sustainable Production of Solketal through the Ketalization of Glycerol

Development of sorption enhanced reactive processes for the sustainable production of Solketal through the ketalization of glycerol

Several research groups around the world have recently identified Solketal as one of the most promising solution towards the valorization of glycerol which has become one of the major topics to ensure the sustainability of the entire biodiesel production chain, since this polyol is obtained as a by-product of the Biodiesel production representing approximately 10 wt% of the total mass of diesel produced. Solketal presents interesting properties as diesel additive which would allow its addition to the final biodiesel formulation not only increasing the quality and the amount of diesel produced but also reducing the overall process by-products formation.

For the synthesis of Solketal, glycerol is commonly reacted with acetone in the presence of an acid catalyst, forming water as by-product [1]. Following the general "Green Chemistry" guidelines, the use of heterogeneous catalysts as ion-exchange resins, zeolites and some silica base materials has been proposed [2]; however, most of the research published focused in batch reactor studies for the optimization of the reaction conditions. In another approach, multifunctional reactors have also been proposed for the production of solketal, which can represent a sustainable alternative for the production of this compound, due to their ability to overcome the thermodynamic limitations associated with this chemical reaction by continuously removing one of the products from the reaction media, which will also result in an increase of the process productivity and in a reduction of its capital cost. In fact, Clarkson et al. [3] and Roldan et al. [4] have proposed the use of reactive distillation and membrane reactors for this purpose, respectively. Although the results presented where satisfactory in terms of glycerol conversion and solketal yield (typically above 90%), it is commonly acknowledged that distillation processes present a high energy demand and zeolite membranes lack long term stability. In this context, the present work represents an innovative approach for the production of solketal making use of sorption enhanced reactive processes, which can outcome the previously studied multifunctional reactors performances based on the high purities and productivities attained by adsorptive-reactors in similar systems [5] and by the process energy efficiency and lower operating temperatures.

A previous study concluded that Amberlyst-35 and ethanol were the most suitable stationary and mobile phases for the synthesis of solketal through a sorption enhanced reactive process, respectively, and fundamental adsorption and reaction data were determined through independent set of experiments providing essential knowledge for the definition of some of the most relevant process operating conditions.

The synthesis of solketal was then experimentally performed in a fixed bed adsorptive reactor. The maximum conversion achieved in a transient state by feeding an equimolar mixture of glycerol and acetone diluted in 50% of ethanol (at 313 K and 5 mL/min) was 60%, while the steady state conversion achieved under these conditions was only 44%. Nevertheless, this value is still close to the reaction equilibrium conversion which is approximately 50%. By changing the reactants molar ratio in the feed mixture from 1:1 to 2:1 (with an excess of acetone) the maximum conversion attained reached 80% in a transient state and the equilibrium conversion was achieved at steady state. These results demonstrate the potential of sorption enhanced reactive processes for the synthesis of solketal. Moreover, it was possible to develop a mathematical model that was able to accurately describe the experimental results obtained in the fixed bed adsorptive reactor experiments considering isotherm operation and an axially dispersed plug flow at a constant bed length and porosity. The non-idealities of the mixture were accounted for by determining the activity coefficients through the UNIFAC method and the internal and external mass transfer coefficients between the liquid and the solid phase were estimated and lumped in a global mass transfer coefficient.

Nevertheless, the fixed bed adsorptive reactor productivity is known to be limited requiring, at the same time, the use of significant amounts of solvent for regeneration purposes, and it would not allow the implementation of a continuous production process. Therefore, the production of solketal in a Simulated Moving Bed Reactor (SMBR) was studied, as this chromatographic reactor overcomes all the aforementioned drawbacks. For that purpose, glycerol and acetone were used as reactants and were introduced in the unit through the feed stream while ethanol was used as desorbent. As solketal is the less retained species it will be collected in the raffinate stream while water, the most retained one, will be collected in the extract stream. Consequently, water is selectively removed from the reaction media leading to the displacement of the thermodynamic reaction equilibrium and considerably increasing the reaction conversion. To evaluate the performance of this technology, a comprehensive simulation study was performed, based on the True Moving Bed Reactor model approach reported elsewhere [6], to optimize the design variables associated with the SMBR operation following to main approaches: using a conventional 4 sections SMBR and feeding a 1:1 mixture of glycerol and acetone, and using a non-conventional 5-sections SMBR in which pure glycerol and pure acetone are fed through independent streams at different columns of the unit (typically feeding acetone closer to the extract port and glycerol closer to the raffinate port to promote the counter-current movement of both of these species based on their adsorption selectivity).

The results for the conventional 4-sections SMBR demonstrated that the purity (in an eluent-free basis) attained in the raffinate stream was 90% which is relatively low compared with the values obtained for similar acetalization reactions performed in SMBR (which are typically comprised between 95% and 99% [5]). The low equilibrium conversion together with the low adsorption selectivity between acetone and solketal when using Amberlyst-35 as catalyst/adsorbent and the significant mass transfer resistances associated with this system represent the most likely causes behind the technology limited performance. On the other hand, the 5-section SMBR unit was able to achieve a solketal purity of 95% without compromising the productivity of the process and without leading to any increase in the desorbent consumption. The main factor contributing for this result was the counter-current contact between the two reactants promoted by feeding them independently at different locations within the unit. As the more retained reactant (glycerol) was fed closer to the raffinate port and the less retained one (acetone) was fed closer to the extract port, a considerable increase in the overall reaction rate and reactants conversion was observed throughout the unit. Additionally, by feeding acetone to a column placed at a larger distance from the raffinate collection port, it was possible to enhance the separation between solketal and acetone.

Therefore, it is possible to conclude that sorption enhanced reactive processes represent a potential alternative for the sustainable production of solketal using glycerol as raw material.

References

[1] Bruchmann, B., Haberle, K., Gruner, H. and Hirn, M., Preparation of cyclic acetals or ketals, US5917059 A, 1999.

[2] Trifoi, A.R., Agachi, P.S. and Pap, T., Glycerol acetals and ketals as possible diesel additives. A review of their synthesis protocols. Renewable and Sustainable Energy Reviews, 2016. 62: p. 804-814.

[3] Clarkson, J.S., Walker, A.J. and Wood, M.A., Continuous Reactor Technology for Ketal Formation: An Improved Synthesis of Solketal. Organic Process Research & Development, 2001. 5(6): p. 630-635.

[4] Roldán, L., Mallada, R., Fraile, J.M., Mayoral, J.A. and Menéndez, M., Glycerol upgrading by ketalization in a zeolite membrane reactor. Asia-Pacific Journal of Chemical Engineering, 2009. 4(3): p. 279-284.

[5] Rodrigues, A.E., Pereira, C.S.M. and Santos, J.C., Chromatographic Reactors. Chemical Engineering & Technology, 2012. 35(7): p. 1171-1183.

[6] Rodrigues, A.E., Pereira, C.S.M., Minceva, M., Pais, L.S., Ribeiro, A.M., Ribeiro, A., Silva, M., Graça, N. and Santos, J.C., Simulated Moving Bed Technology: Principles, Design and Process Applications. 2015: Elsevier Science.

578342 Gradual Synthesis of Heat Exchanger Networks Under Economic, Environmental and Safety Considerations

The heating and cooling needs in the chemical process industry are typically covered using external utilities. Process intensification in the form of energy integration among process streams is an effective tool to achieve more energy-efficient designs. We address in this work the design of heat exchanger networks (HENs), but instead of developing a design aiming for a full energy integration we examine the effect of performing gradual process integration by adding process-process exchangers until the complete integration is achieved to analyze the step-by-step effect of such a strategy.

The problem starts by considering m hot and n cold process streams to be integrated, for which targets for minimum energy consumption are determined through pinch point techniques. Then, the individual contribution of each heat exchanger used to integrate energy is analyzed, and the compromise between the investment cost and the reduction of utility consumption is evaluated. To complement the analysis environmental and safety factors are considered for each of the networks that result from a structure with no energy integration to a final one that achieves complete integration that meets the minimum energy consumption targets.

The strategy for gradual energy intensification is based on the heat duties required by each exchanger. Therefore, we start with the implementation of the heat exchanger that shows the largest duty, for which its size, cost, and contribution towards the objective of minimum energy consumption are determined. The effect on the environmental and safety indicators is also assessed. This procedure is applied sequentially until full energy integration is accomplished.

The results are analyzed through several metrics. From economic terms, each integrated structure is evaluated taking into account the saving generated through energy integration and the investment cost for the heat exchanger(s). If marginal return on investment is of interest, it is shown that partial integration provides a better policy. Environmental and safety indicators are also considered. The environmental impact of each partial network is evaluated in terms of the reduction of CO_2 equivalent emissions that are obtained as heat exchangers are added to the network. Safety aspects are evaluated through the Inherent Safety Index for Shell and Tube Heat Exchangers (ISISTHE)[1], which rates inherent safety considering pressure, heating value, combustibility and an adjusted temperature difference.

The procedure is applied to several case studies ranging from systems with few streams to the aromatics plant problem.[2] By considering the three factors (economic, environmental and inherent safety) it is shown how one can identify the level of integration at which an optimum point that meets the best compromise among them is obtained.

[1]Pasha, M.; Zaini, D.; Shariff, A. M. Inherent safer design for heat exchanger network. J. of Loss Prevention in the Process Industries. 2017, 48, 55-70.

[2]Linnhoff, B.; Ahmad, S. Cost optimum heat exchanger networks-1. Minimum energy and capital using simple models for capital cost. Computers Chem. Engng. 1990, 14, 7, 729-750.

571849 Rapid Risk Assessment and Process Optimization to Enable FIH

Given the competitive landscape faced by the pharmaceutical industry, rapid delivery of first in human (FIH) clinical materials through efficient risk identification and process development is critical. Identifying important API quality attributes and major liabilities that may affect them focuses the development of an enabling process. This presentation provides a case study on optimization of a complex, fit-for-purpose process to enable scale-up for an FIH delivery of a small molecule API for parenteral use. Some of the challenges and risks inherent to the process included the use of hazardous hydrogen fluoride chemistry, unit operations such as prep-HPLC, ion-exchange, and lyophilization, specific requirements for API material properties, and stringent bioburden and endotoxin control. Liabilities identified included residual fluoride impacting the API stability, residual modifier from the prep-HPLC mobile phase impacting potency, water-uptake of API due to its hygroscopicity, and variability in the counterion content of the API. A control strategy was devised and implemented across the reaction, crude precipitation, prep-HPLC, concentration, ion-exchange and lyophilization to address these liabilities. Lastly, a risk assessment for bioburden and endotoxin was carried out, and specific processing requirements were outlined in order to ensure adequate control. The identification and mitigation of major risks through targeted process development resulted in successful scale-up production of the drug substance.

576567 Optimizing Continuous Twin-Screw Wet Granulation for a Batch to Continuous Conversion

Optimizing Continuous Twin-Screw Wet Granulation for a Batch to Continuous Conversion

Qiushi Zhou,

Fernando Muzzio

Department of Chemical and Biochemical Engineering, Rutgers, The State University of New Jersey, 98 Brett Road, Piscataway, New Jersey 08854-8058, United States

Continuous manufacturing has advantages in increasing process efficiency, product quality, and reducing operating cost comparing to the traditional batch manufacturing process. Continuous wet granulation is one of the most important methods that has been widely used among downstream pharmaceutical manufacturing. This project shows the feasibly of converting a high-shear batch granulation to a continuous twin-screw granulation method. Particle sizes of granules can be predicted by the model generated, resulting in particle sizes that match the high-shear batch granulation process.

To determine the ideal design space of the twin-screw granulator, a preliminary design space study was used to define the range of working conditions. After obtaining the condition range, a face centered cubic design was used in the focused design space. The particle size was the key resultant factor that was monitored. The results from the DoE showed that screw speed had less influence on the particle size than both the L/S ratio and alcohol content. The results suggested that a clear trend in particle size was observed for L/S ratio and granulation fluid composition. Particle size decreased with an increase in alcohol content in the granulation fluid and a decrease in the L/S ratio. Combining the data from the preliminary design space study and designed experiments enable the determination of a stepwise regression that was able to predict granules properties.

567903 Regeneration Scale-up Methodology Applied to Bio-Based Isobutanol Dehydration to Linear Butenes

The technology of isobutanol dehydration to linear butenes has been developed in a joint collaboration of IFP Energies Nouvelles (France), Total Research & Technology Feluy (Belgium) and Axens (France). The dehydration reaction takes place in gas phase with a low water dilution of the alcohol feedstock and is carried out in a fixed-bed multitubular reactor loaded with a ferrierite (FER) zeolite-based catalyst. The process has high isobutanol-to-olefins conversion and yield and strong selectivity to linear butenes. One of the process drawbacks is the progressive catalyst deactivation due to coke formation and the respective conversion loss at the end of the catalytic cycle. Therefore, the process operates in cyclic mode that alternates highly endothermic reaction and highly exothermic regeneration steps. The proposed presentation describes the step-by-step regeneration scale-up methodology to reduce the risk involved in such process scale-up. Although main design criteria are induced by the endothermic dehydration in order to meet target production, validation of the adequacy of such scale-up with exothermic regeneration is necessary. Thus, a sufficient knowledge and control of the various chemical and physical phenomena occurring during regeneration were acquired. Combustion reactions scheme was set-up thanks to controlled thermogravimetric experiments at micro scale and then directly validated by dynamic simulation on pilot unit taking into account heat transfer phenomena. At last, a specific sizing and rating of process equipment in non-steady mode was realized to optimize and ensure control of industrial regeneration protocol. This methodology can be generalized for a wide variety of chemical processes using heterogeneous catalyst.

The first step of the regeneration scale-up methodology is the establishment of a regeneration protocol. Catalytic tests have been carried out in a packed bed tubular pilot plant unit consisting of a tubular reactor filled with catalyst. The regeneration gas temperature has been maintained by heating the reactor shell with an electrical furnace. The major challenge to overcome in the case of FER-based catalyst is its high sensitivity to regeneration and relatively low capacity to be regenerated. After series of catalytic tests, we have observed that the optimal composition of the gas for FER-based catalyst regeneration comprises a variable mixture of nitrogen and oxygen. Oxygen content varied during regeneration, starting from relatively low (<1% mol.) and ending at relatively high value (<10% mol.). The temperature of the regeneration follows a slope-plateau program. It is increased at the end of protocol in order to complete the coke calcination. Moreover, we have found out that the presence of a relatively low amount of water in the regeneration gas provokes performances degradation. The regeneration gas should not contain any residual free water for better overall catalyst life duration. As a result, the regeneration protocol is an operational program that includes a variation of gas composition and temperature.

The regeneration protocol described here above has been optimized and tested at the pilot unit after several catalytic runs with various grades of isobutanol. The protocol permits to fully restore the catalytic performance and to preserve catalyst cycle length.

The second step of regeneration scale-up methodology corresponds to the determination of coke composition and reactional kinetics of coke combustion via a thermogravimetric analysis (TGA) of coke, in order to enable control of the exothermic reactions in further extrapolations. Due to this analysis, we have observed that the coke is mainly composed of two large hydrocarbon families that burn at different temperature ranges (see Figure 1).

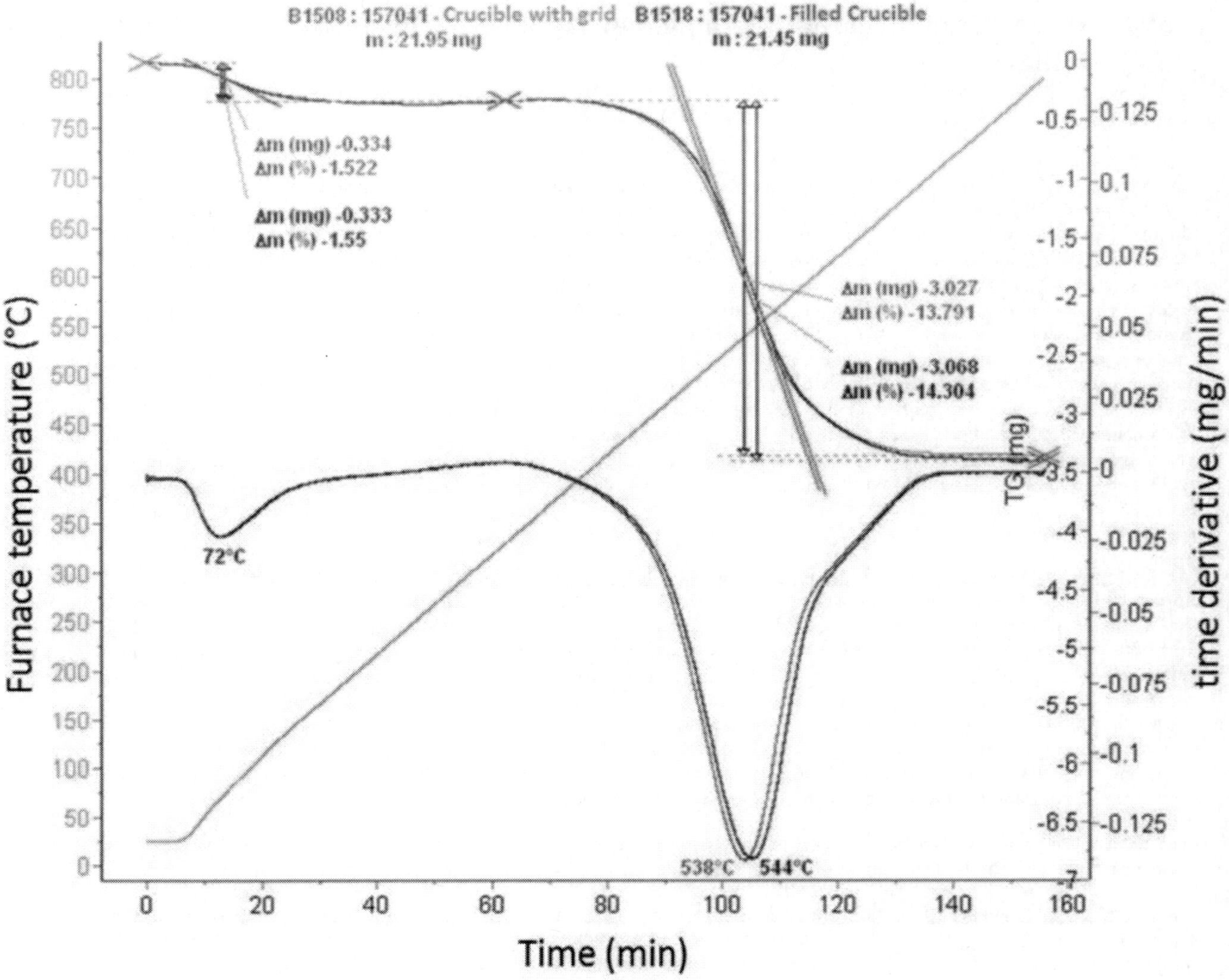

Figure 1. Thermal curves of TGA experiment

Following a weight loss during a thermogravimetric experiment, activation energies of coke combustion via Kissinger-Akahira-Sunose equation have been identified. Results obtained with this equation are coherent with those from bibliographic sources. Moreover, a TGA experiment allows us to determine partial orders of combustion reactions by solving a mass balance of reactions. We have compared mass-loss curves obtained experimentally with those traced with mass balance equation (see Figure 2) and found values of orders of reaction.

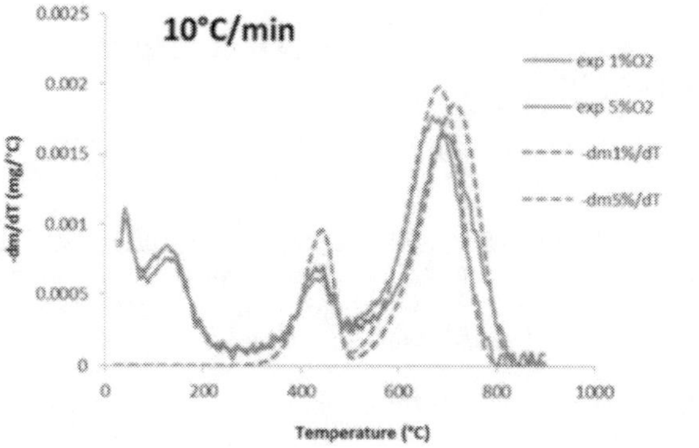

Figure 2. Mass-loss curves of TGA analysis of coke

The third step of the regeneration scale-up methodology is a dynamic simulation of the catalyst regeneration in order to confirm the results obtained experimentally. Reaction kinetic parameters determined by TGA have been implemented in a 1D plug-flow reactor simulator, simulator with 2D heat transfer (radiative and conductive) characterization, in order to model coke combustion in dynamic mode and to trace carbon dioxide/carbon monoxide curves. Two major hydrocarbon families identified by TGA tests have been represented by mono-aromatic and poly-aromatic molecules. This simplification does not affect the accuracy of the results and is globally very robust: curves obtained with the simulator are coherent with experimental curves. This fact confirms a correctness of dynamic regeneration modeling (see Figure 3). The simulator allows to extract a dynamic mass balance data of coke combustion.

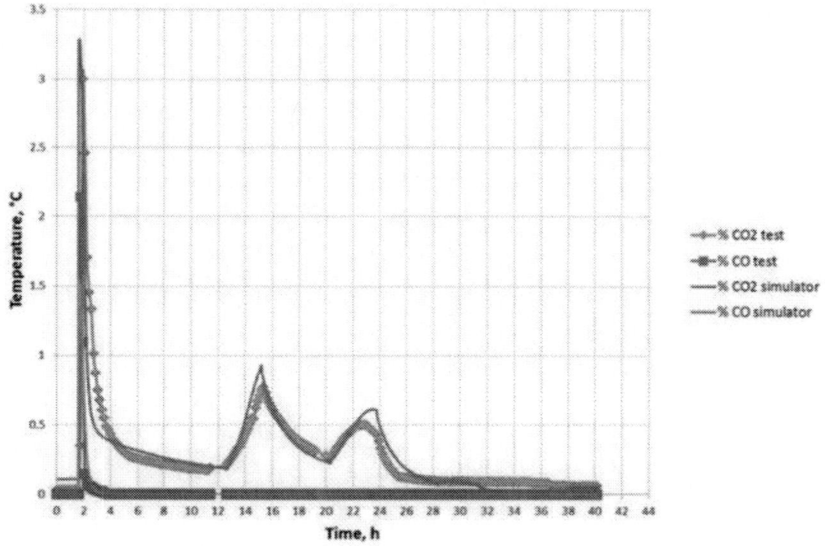

Figure 3. Comparison of carbon dioxide/carbon monoxide curves obtained experimentally and with simulator

Then, the following hypotheses for process extrapolation to an industrial scale have been made:

- Mass balance data obtained with the dynamic simulator of the pilot unit is identical for an industrial unit
- Mass of coke per mass of catalyst of the pilot unit is identical for an industrial unit

- Mass of catalyst required for an industrial unit is calculated conserving the same Weight Hourly Space Velocity (WHSV) as for the pilot unit

The fourth step of the regeneration scale-up methodology corresponds to transfer of dynamic mass balance data to static simulation in order to determine heats of coke combustion. Firstly, we have calculated the quantity of coke on the catalyst of an industrial unit. Secondly, the regeneration timeline has been divided into several phases. We have calculated the quantity of coke burnt at each phase and further determined a mass flowrate of coke per hour. Lastly, knowing the composition of coke burnt, heats of combustion and quantities of regeneration gas have been identified.

The last step of the regeneration scale-up methodology is the sizing and rating of process equipment in non-steady modes. Data obtained in the previous step have been used in order to fix a maximum quantity of regeneration gas required. It is important to mention that the multitubular reactor, initially designed to perform an endothermic dehydration reaction, also allows an efficient heat extraction from the exothermic combustion reaction during the regeneration phase.

In conclusion, the above described methodology allows to develop a process of effective FER-based catalyst regeneration in a multitubular reactor. This methodology can be applied for other technologies employing a heterogeneous catalyst that requires periodic regeneration due to coke formation.

571083 Design and Implementation of Robust Multicomponent Working Fluid for Organic Rankine Cycle

The organic Rankine cycle (ORC) has been actively developed as a way to generate electricity, using waste or renewable heat resources. Many researches are concentrated on the study to enhance the performance of the working fluid for ORC as it is considered as the most important factor for the overall efficiency of ORC. The design of a working fluid is highly dependent on the characteristics of heat sinks or sources. One common issue on designing ORC utilizing heat sources or sinks, which is not isothermal in heat exchangers, is that it diminishes thermal efficiency by increasing the amount of exergy destruction in heat exchangers. Many studies have shown that it is inevitable to use a multicomponent working fluid to achieve high thermal efficiency [1-3]. This is because, with optimally selected working fluid substances and composition, using a multicomponent working fluid is advantageous in minimizing the exergy destruction in the heat exchanger, since it better fits the temperature glide of the heat source or sink, compared to single component working fluids. In this context, many studies have been performed to develop the method for designing the optimal working fluid mixture for ORC [1-3]. However, although the previous methods for designing the working fluid have succeeded in significantly improving the efficiency of ORCs, the multicomponent working fluid obtained as a result of the designing methods would not be suitable for an actual ORC operation due to its vulnerability to uncertainties in operating conditions and thermodynamic parameters.

Only a few studies have been performed to address the ORC design problem considering uncertainties [4-6] but these studies have common limitations to design robust working fluids. They used Monte Carlo (MC) simulations for evaluating the effect of uncertainties on ORC performance. MC simulations can provide the statistical moments required for stochastic optimization, but its slow convergence rate precludes the optimization from obtaining the robust design in a reasonable amount of time. Also, none of the above studies have considered the uncertainties in the composition of multicomponent working fluid, which can vary due to its thermal degradation or leakage [7].

In this study, a systematic method to design a robust ORC using LNG and ternary working fluid, which yields maximum power output even when the composition of the working fluid and the temperature of the heat source vary from the nominal points during operation of ORC and the values for thermodynamic parameters have ranges has been developed. Since it is computationally burdensome to consider all the uncertainties at once, the proposed method initially seeks the optimal combination for the working fluid mixture by considering only uncertainty in composition. It seeks the optimal composition and ORC operating condition giving the maximum weighted sum of the mean and the variance of the ORC power output. To suppress the factors that adversely affect the operation of ORC (violation of minimum temperature difference in heat exchanger and formation of liquid droplet in expander), the objective function is penalized when they occur. The procedure to derive the statistical moments consists of two steps. Initially, the required heat exchanger area is obtained by simulation of ORC model with a nominal operating conditions (composition, pump discharge pressure, and expander discharge pressure). At the next step, the simulation is carried out again with the obtained area and the varying composition and parameters. The mass fraction of each substance in the working fluid is assumed to follow uniform distribution centered at the nominal point to consider the worst case of ORC operation. To obtain the mean and variance with a small number of simulations, Polynomial Chaos Expansion (PCE) with sparse grid quadrature is employed. It has been shown that small changes in composition and parameters can have serious consequences for stable operation of ORC, and the design of working fluid by the proposed method allows flexible ORC operation despite the existence of uncertainty in the composition [8].

Finally, the proposed design strategy takes into account uncertainties in thermodynamic parameters and heat source in addition to composition. Using the selected working fluid which was turned out to be the most insensitive from the uncertainty of composition, optimization is carried out again when the critical temperature and pressure of each substance composing the working fluid vary within their measurement uncertainty, which can be found in the literature [9]. Also, the uncertainty in the temperature of the heat source is taken into account for further enhancing the operational flexibility of ORC. In sum, since the

ORC uses ternary mixture as its working fluid, the design of ORC was performed assuming a total of the nine parameters or design variables under uncertainty (2 composition variables, critical temperature and pressure for each of the working fluid substance, and the temperature of the heat source), which requires excessive amount of computation using the method used when considering only the composition uncertainty. Therefore, the optimization using a surrogate model is devised to efficiently find the optimal and robust ORC design. Because the proposed surrogate model is constructed based on PCE, the statistical moments can be derived analytically, which leads to reduce the time for optimization drastically. Comparing the ORC, which was designed under more uncertainties, to the design obtained under only the composition uncertainty, the former design showed higher output even when the parameters and design variables varied from the nominal points.

[1] Lee, U.; Kim, K.; Han, C. Design and optimization of multicomponent organic rankine cycle using liquefied natural gas cryogenic exergy. Energy 2014, 77, 520–532.

[2] Kim, K.; Lee, U.; Kim, C.; Han, C. Design and optimization of cascade organic Rankine cycle for recovering cryogenic energy from liquefied natural gas using binary working fluid. Energy 2015, 88, 304–313.

[3] Liu, Y.; Guo, K. A novel cryogenic power cycle for LNG cold energy recovery. Energy 2011, 36, 2828–2833.

[4] Frutiger, J.; Andreasen, J.; Liu, W.; Spliethoff, H.; Haglind, F.; Abildskov, J.; Sin, G. Working fluid selection for organic Rankine cycles–Impact of uncertainty of fluid properties. Energy 2016, 109, 987–997.

[5] Mavrou, P.; Papadopoulos, A. I.; Seferlis, P.; Linke, P.; Voutetakis, S. Selection of working fluid mixtures for flexible Organic Rankine Cycles under operating variability through a systematic nonlinear sensitivity analysis approach. Appl. Therm. Eng. 2015, 89, 1054–1067.

[6] Bufi, E. A.; Camporeale, S. M.; Cinnella, P. Robust optimization of an Organic Rankine Cycle for heavy duty engine waste heat recovery. Energy Procedia 2017, 129, 66–73.

[7] Kruse, H.; Rinne, F. Performance and leakage investigations of refrigeration and airconditioning systems using refrigerant mixtures as working fluids. Int. Refrig. Air Condition. Conf. 1992, 621–630.

[8] Kim, K.; Kim, J.; Kim, C.; Lee, Y.; Lee, WB. Robust design of multicomponent working fluid for organic Rankine cycle. Ind. Eng. Chem. Res. 2019, 58, 10, 4154-4167.

[9] Chase Jr, M. W. and Tables, N.-J. T. Data reported in nist standard reference database 69, june 2005 release: Nist chemistry webbook. J. Phys. Chem. Ref. Data, Monograph, vol. 9, pp. 1–1951, 1998.

580165 Study on Ozone Formation Characterization during the Flaring of Chemical Plant Shutdown for Proactive Flare Minimization and Air-Quality Control

Flaring is a necessary measure for chemical plant safety. However, the industrial flaring generates large amounts of VOCs and NOx, which could transiently aggravate regional ozone concentrations in the presence of sunlight that could affect on human health and negative societal impacts. Note that VOCs from the flaring of chemical plant shutdown mainly consist of ethylene, ethane, propylene, propane and other hydrocarbons, which have the different potential for ozone formation. So, it is important to know that the sensitivity and contribution of the regional ozone impact due to the different chemical species from the flaring of chemical plant shutdown. In this paper, plant-wide dynamic simulations for plant shutdown flare emissions with regional air-quality modeling were coupled together to quantify the air-quality impact (i.e., ozone concentration) due to the different chemical species. Also, deep insights of the ozone formation characterization are provided for proactive air-quality control and optimal flare minimization. Case studies indicate that ethylene and propylene are main species for ozone formation during the flaring of chemical plant shutdown, which accounts for 58.8 % and 23.1 % ozone formation potential, respectively. Thus, it becomes more important and attractive for the industry from the viewpoint of ozone pollution control to make flare minimization strategies by recycling the ethylene and propylene instead of all chemicals, which can save lots of resources and achieve the cost-effective strategies for air-quality control. This study could provide valuable and quantitative support for both the industry and environmental agencies on developing proactive and optimal flare minimization strategies in the future.

562851 CFD-FEA Analysis for Scale-up of a Twin Screw Extruder Kneading Block Section

The kneading block sections of co-rotating twin screw extruders induce both distributive and dispersive mixing of the components of the fluid stream as the material is sheared and elongated by the clearances between the intermeshing kneading blocks and the barrel walls. Narrow kneading discs tend to provide more dispersive (and consequently distributive) mixing, and are the focus of this study. In practice, the fluid stream will often be a multi-phase mixture where the primary phase is a molten polymer and the added phases may be solid particles, or immiscible fluids. In the current investigation, the fluid steam mixture will be treated as pseudo-homogeneous, high-viscosity, shear-thinning material (i.e., polymeric melt). Neutral (non-forwarding) kneading blocks can be treated as fluid filled, which is the case for the current study.

Although the mixing performance is of obvious importance, this study is focused on the mechanical integrity of the narrow kneading blocks under various operating conditions. If the fluid forces acting on the kneading block lobes is too high, the lobes can bend, twist, or break causing significant damage to the twin screw extruder at significant cost of both capital and production capacity.

This study evaluated multiple methods to simulate the flow and pressure characteristics of a polymeric fluid in the intermeshing co-rotating system at different geometric scales (and resulting disc cantilever lengths) from development scale to production plant. The resulting fluid pressure contours on the kneading discs were applied to the disc surfaces within a finite element analysis to evaluate the structural deformation and overall structural integrity of these kneading elements (by comparing the resulting stresses with the yield stress of the disc material). If the risk of damage is determined to be high, this analysis approach can be used to modify the kneading block design to withstand the anticipated stresses.

566261 Investigation of Soil Corrosive Effect on Buried Gas Pipeline: New Correlation Development

One important category of transportation infrastructure is underground pipelines. Corrosion of theseburied pipeline may cause pipeline failures with the attendant hazards of property loss andfatalities. Therefore, developing the capability to estimate the soil corrosivity is important for designing and preserving materials. In this work, corrosion behavior of pipelines buried in soil for Fars province in Iran was investigated. Influence of parameters such as electrical resistivity, pH, structure and composition of soil on the corrosion behavior of buried pipeline were studied. Thus, a new standard criteria methodology has been proposed, illustrated, and validated for the analysis of external corrosion of buried pipelines which they were coated by three type of coating e.g., Coal-tar, three-layer polyethylene and Bituseal. It was noted that this method for soil corrosivity is directly based on Dechema lab procedure. Results indicated that the Bituseal enamel was more resistive than Coal-tar and 3LPE coatings so that there was no change on the current density when Bituseal enamel had been used. On the other hand, the Coal-tar was lower corrosive resistivity than others due to existence of more porosity in its structure. As the same way, the current density consumption and its gradient by changing soil parameters were higher than 3LPE and Bituseal enamel coatings.

574862 Constrained Optimisation of Cell Culture Feeding Strategy and Temperature Shift Duration to Enhance Monoclonal Antibody Titre and Purity

Recombinant proteins have been extensively studied for their wide therapeutic and research applications. The main therapeutic product category is that of monoclonal antibodies, which have been approved to treat a variety of chronic and life-threatening diseases. Increasing production titre has been achieved mainly by cell culture medium improvement and genetic engineering. However, mathematical modeling of bioprocess dynamics is a valuable tool to improve industrial production at a much faster rate and lower cost in comparison to statistical design of experiment approaches. A holistic insight to comprehend the interplay between crucial processes variables can be gained through the application of process system engineering tools. In this study, a single stage population balance model has been built to capture mammalian cells behaviour in fed-batch bioreactors. The model represents two operating modes; the first at constant physiological temperature and the second with a shift to mild hypothermic conditions (32°C) from the fifth day onwards, which represents common industrial practice.

The model considers the dynamic profile of substrates and metabolites, product titre and one of the main cell-derived impurities, host cell proteins (HCPs), which can impact product immunogenicity and integrity. Certain HCPs are naturally secreted, while others are released as a result of cell death and lysis. Our model considers these three sources of HCPs to determine their overall concentration at harvest. Culture osmolality is also considered as a determining factor for cell growth rate and size distribution.

The model parameters have been estimated with high confidence using a recently published dataset for antibody-producing Chinese hamster ovary (CHO) fed-batch cell cultures for physiological temperature and with a shift to mild hypothermia in mid-exponential growth phase. The model was then used to optimise feeding volume and frequency as well as duration of temperature downshift subject to maintaining culture viability above 80%, as is current industrial practice. and this has been set as a constraint in all our model optimisation runs. Specifically, we looked at three scenarios: (a) increased feeding to sustain high viability for longer, which maintained titre but reduced cell death; (b) a 50% reduction in feed volume to avoid overfeeding, which led to a slightly higher titre compared to the control; (c) a shift in culture temperature on day 6 which gives the highest product/ HCP ratio.

In general, higher product titres and prolonged culture viability can be attained at the expense of higher feeding volume. However, when a constraint on HCP concentration is also applied model-based optimisation results in shorter culture duration and, in turn, overall lower antibody titre. This study shows the usefulness of mathematical modeling for exploring trade-offs in bioprocess performance. It is also a viable tool to systematically accelerate the duration of processes development and optimisation. Integrating this model with a downstream purification model to evaluate the cost of removing these fractions of impurities, can help determine what concentration of HCPs can be economically tolerated in the cell culture supernatant and aid whole bioprocess design.

564778 Control of Glycosylation and Titer in Fed-Batch Monoclonal Antibody Production

Therapeutic monoclonal antibodies (mAbs) are currently produced commercially in fed-batch cultures using media recipes and protocols based primarily on heuristics. The primary production objectives are maximizing mAb titer (productivity requirement) and achieving desired glycan distributions (quality attribute) that result from glycosylation—an enzymatic, post-translational process that attaches sugar molecules (glycans) to antibodies. The distribution of glycans is a critical quality attribute of mAb manufacturing, because it can either promote or inhibit desirable therapeutic effects of mAb drugs. However, controlling glycosylation is challenging primarily because it is a complex, non-template driven process that takes place within the cell. Consequently, in addition to the apparent chaotic nature of glycan attachment, the variables available for influencing glycosylation are far fewer than the controlled variables. Furthermore, as a consequence of the multifaceted nature of glycosylation, the process is difficult to model, and on-line sensors do not currently exist for characterizing glycan distribution frequently enough for effective on-line control. Nevertheless, recent advances in both glycosylation modeling and glycan assays have improved the possibility of implementing model-based control of glycosylation in place of the current heuristics-based approach.

Here we present an overall model-based strategy for meeting productivity and product quality requirements in the manufacture of mAbs. The proposed multivariate cascade control system consists of two nested control loops. The outer loop, which is responsible for controlling product attributes such as mAb titer, cell density, and the relative abundance of dominant glycans, uses a model predictive controller (MPC) to determine the setpoints for the inner loop. The inner loop in turn employs multiple PI controllers to sustain the cell culture by maintaining the levels of nutrients, pH, temperature, etc., at desired values. Because measurements of glycan distribution are not available frequently enough, the outer loop incorporates a state observer which uses available sensor measurements and model predictions to reconstruct the complete system state that the MPC uses, along with the process model, to determine the desired control policies.

As a first step toward implementing on-line control of industrial fed-batch mAb manufacturing processes, we tested the complete control system in a MATLAB simulation, investigating a variety of industrially relevant process conditions under several operating sensor configurations—frequencies, qualities, and delays. The simulation results offer insight into optimal sensing strategy as well as the usefulness and limitations of the proposed control system.

576555 Using High-Throughput Microbioreactors to Fit Raman Spectroscopy Model for Measuring Key Metabolites in Monoclonal Antibody Cell Culture Manufacturing Process

Raman spectroscopy is a process analytical technology for simultaneous in-line measurement of important process outputs in therapeutic protein production via mammalian cell culture, such as concentrations of key metabolites or even the protein being expressed. However, due to the complex and shifting nature of the various chemical species present in the cell culture system, Raman models will likely need to be at least partially trained on a process-specific basis. They must also be robust enough to maintain accuracy throughout the expected manufacturing process operational ranges. This presentation is a proof-of-concept study to demonstrate a reliable Raman model can be developed in parallel with standard process development experimentation for later deployment in a scaled-up manufacturing process.

Sartorius ambr® 250 high-throughput microbioreactors were used for cell culture process development studies for a CHO cell line expressing an IgG4 monoclonal antibody ("mAb1"). Process parameters including initial cell density, nutrient feed strategy, pH, and dissolved oxygen control for a fourteen-day bioprocess were examined, and standard off-line monitoring of all process outputs were carried out. Concurrently during this high-throughput study, Raman spectra were also periodically generated from bioreactor samples both before and after being spiked with known concentrations of key metabolites. These data were used to train a Raman model and verified against similar spectra obtained from larger bioreactors.

In conclusion, this study generated a scalable Raman spectroscopy model for various metabolites that can be used for monoclonal antibody process control and performance monitoring. A robust model can be trained over a wide range of processing conditions, in a manner that would not interfere with standard process development or require additional experiments. The comparison of the outputs from this Raman model to the standard off-line data will also be presented.

567854 Evaluation of Process Parameters Influencing Immobilized Metal Affinity Chromatography

The production of biopharmaceutical drugs requires high yielding and robust purification methods. Affinity chromatography is a preferred method to capture product from culture media due to its high selectivity. However, many protein therapeutics lack a suitable affinity capture modality and therefore require use of engineered tags for affinity purification. Immobilized metal ion affinity chromatography (IMAC) has been a popular choice for the purification of histidine-tagged molecules. One drawback of this technology is that process performance is often variable, making it difficult to predictably scale-up. This study contributes to the understanding of process parameters impacting performance and how to optimally control these parameters. Multivariate experiments were performed to investigate the binding of histidine-tagged proteins to commercial metal chelating resins. This study revealed that imidazole concentration, load capacity and metal binding ability of the resin influenced the protein binding capacity. This study also examined the impact of histidine tag structure on chromatographic performance. Adsorption isotherms were constructed to elucidate differences in equilibrium binding of multi-histidine tagged material. A mathematical model was then employed to describe the impact of changing process parameters on product quality. The findings of this study are useful to understanding factors impacting IMAC operation, including the relationship between process parameters, tag structure, process performance, and product quality.

566098 Investigation of Bio-Capacitance Driven Process Variation in a Site to Site Large Scale Technology Transfer Campaign

Bio-capacitance in cell culture processing allows for more automated process controls and on-line biomass monitoring. Here we presented a case study where the large-scale manufacture process (15,000L) is monitored and nutrient-fed based on the dual bio-capacitance automation driven protocol. In one of the batches, unexpected process variation was observed. Higher amount of feed, out of trend nutrient concentration and accumulative waste level were observed during the process which can be a signature of nutrient overfeed. Our investigation suggested that the primary and backup bio-capacitance reading have significant deviation especially at the earlier days. When bio-capacitance reading deviation happened, viable cell volume (VCV), calculated by Viable cell density (VCD) and cell diameter, was used to justify which probe signal to use. This is because capacitive signal is to measure the dielectric properties in the membrane of viable cell. Historical VCV/bio-capacitance ratio is referenced as process control strategy to translate the online bio-capacitance reading and offline Vi-cell measurement. In conclusion, the variation of Vi-cells to Vi-cells in addition to deviation of dual capacitance readings, created the signal to demand for higher feed volume. As a result, it might cause negative impact on the product quality such as glycosylation. Based on this large-scale process experience, we developed a more robust bio-capacitance control strategies to facilitate the site-to-site process transfer in the future.

566086 Characterization of Rotary Lobe Pump Damage in Large-Scale Cell Culture Perfusion Process

Perfusion bioprocessing is becoming more prevalent for increasing productivity and shortening seed train and bioreactor stages. When implementing the new perfusion platform in N-1 large-scale manufacturing process (3750L), unexpected viability drop was observed at day 1. This was not seen at the bench scale satellite runs using the same culture and medium. Higher turbidity and lactate dehydrogenase (LDH) were in the large-scale bioreactor also confirmed the cells were stressed more significantly. Towards the end of the process, the transmembrane pressure (TMP) increased significantly and fouled the hollow fiber membrane. Several investigations were conducted to understand the potential root cause including bioreactor sparging, hollow fiber filter shear, the piping of skid design and operation of the rotary lobe pump. A quick mitigation is to lower RPM in the rotary lobe pump which immediately reduced the cell damage. A follow up study to utilize a smaller scale of rotary lobe pump in the lab was conducted. RPM, flow rate, differential pressure across the pump were studied to identify the possible mechanism of cell damage. The results suggested that differential pressure across the pump was the primary cause of cell damage. A small-scale robustness model was developed to better predict manufacturing performance and improve the large-scale operation in the future.

567770 Process Intensification of the Alkoxylation of Fatty Amines

Process Intensification of the Alkoxylation of Fatty Amines

Pia Müller[1*], Wyatt Winkenwerder[2], John van der Schaaf[1]

1 Department of Chemical Engineering and Chemistry, Eindhoven University of Technology, P.O. Box 513, 5600MB, the Netherlands; 2 Nouryon, Brewster, New York 10509, United States

*Corresponding author: p.muller@tue.nl

j.vanderschaaf@tue.nl

wyatt.winkenwerder@nouryon.com

1. Introduction

Alkoxylated fatty amines are widely used as non-ionic surfactants, leading to a steady increase in industrial demand over the past six decades. These compounds are key components in products of numerous industrial branches, e.g. oil drilling, paper manufacturing, textile industry, as well as fiber manufacturing, paints, dyes and plastics [1]. These compounds are typically produced by reacting fatty amines with ethylene oxide (EO) or propylene oxide (PO) in large fed-batch reactors, often reaching the size of 30 m^3. The process is intentionally de-intensified, due to its strongly exothermic nature ($\Delta_R H = -120$ kJ/mole of EO) and the safety risks associated with EO. Therefore, the epoxide is slowly added to the fatty amine to keep the exothermic reaction under control. In the past, attempts have been made to intensify the reaction by using venture-loop or spray-tower reactors [2], [3]. However, due to the required active mixing, these processes necessitate a power input higher than the typical set-up, and fail to provide significantly improved product output. The aim of our work is to transfer this industrially important reaction to using modern technology to a 21st century process. One possible option to be is shown in Figure 1.

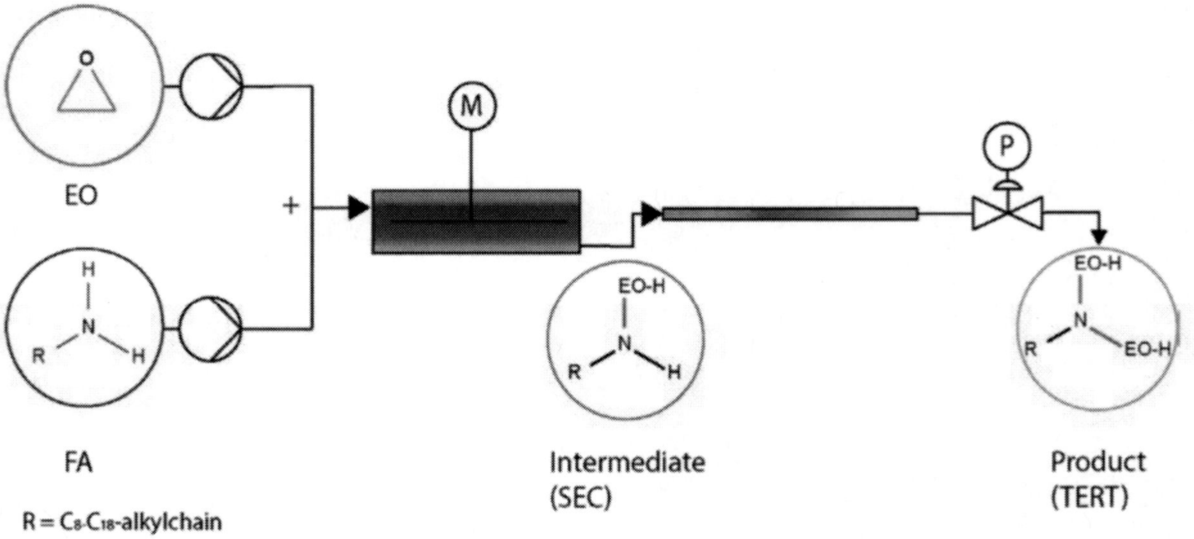

Figure 1: Schematic of possible, intensified reaction setup. In total two mole EO (or other epoxides) react with the fatty amines to

form a di-alkoxylated amine. The process includes a CSTR in series with a plug flow reactor.

This study focuses on gaining detailed kinetic insights and their application to the development of continuous flow systems to improve the reaction efficiency whilst maintaining safety standards.

2. Methods

To achieve our aim a kinetic study was necessary. In a first step we used model compounds such as butylene oxide (BO) or propylene oxide (PO) with octylamine (OA) in a lab scale reactor. A temperature window of 120 °C to 160 °C was explored by analyzing the mixture with GC, NMR and inline FTIR. The process imitated the industrial process by adding the epoxides in a lower molar ratio stepwise to the amine.

In a second steps the collected data were then applied to the design of a flow set-up utilizing microchannels as perfect plug-flow reactors. A comparable flow system was also employed to determine the kinetics of ethylene oxide (EO) with OA over a temperature range between 120 and 170 °C. For safety reasons and accurate contact times it was important to work in a fully liquid system. The system was therefore pressurized by a back pressure regulator to 12 bar (PO and BO) or 60 bar (EO). Such pressure for PO was chosen after a large negative deviation from ideal mixing (8 bar measured at 150 °C to 22 bar in mixture) behavior was observed and a NRTL model was designed. For EO a high tolerance of the theoretical vapor pressure was chosen.

Finally, the reactions with PO and OA was transferred to different set-ups of potential industrial interest. Options like spinning disc reactors or other CSTR types were tested, as well as the recycling of an intermediate mixture

3. Results and discussion

For the reaction of BO and PO with octylamine experimental data were collected in semi-batch and flow:

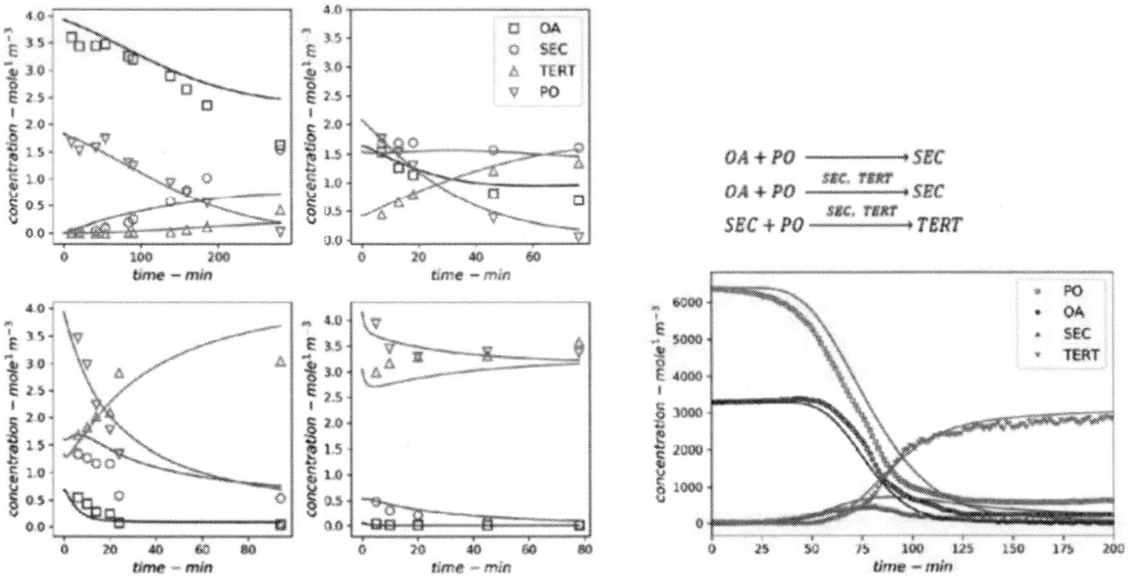

Figure 2: Experimental data in fed-batch (left) was collected over four addition steps of PO of each 0.5 mole per mole OA. In flow (right) the molar ratio initially was 2 mole PO per mole. Both experiments were performed at 130 °C. An auto-catalytic model (upper right) has been used to simulate both experiments.

The reaction is initially very slow in both batch and flow. The reaction accelerates after a certain amount of product is formed, which is typical for auto-catalytic reactions. The advantages of going from fed-batch to flow are clearly visible in Figure 2. What is impossible to achieve in batch, i.e. high conversion in only one step, is feasible in flow. Through the use of microchannels the volume could be decreased from 100 to 7 ml. Thanks to the intensified mixing the reaction time for achieving full conversion could also be decreased by a factor of 1.5-3.0 depending on the epoxide and conditions. Furthermore, the continuous set-up is superior due to the higher saturation of the amine with epoxides achieved under pressure. The overall reaction for PO was 1.5 times as fast as for BO. It is expected for EO an increase in rate by at least a factor 2 compared to PO. The use of a CSTR will lead to an additional decrease of reaction time, due to a better back mixing, which would enhance the inherent auto-catalytic effect of the reaction.

4. Conclusions

Process intensification for the alkoxylation of fatty amines was achieved with a test flow system. A rate increase by a factor 1.5-3.0 could be reached for the reaction of propylene and butylene oxides with the model substrate octylamine. Using a continuous microchannel flow system, the intrinsic safety risks associated with the process could be overcome. A kinetic model that fits the experimental data for the reactions of BO, PO, and EO with amines was found. Using this model, different process configuration options are easily compared for performance.

References

[1] T. M. Schmitt, *Analysis of Surfactants, Second Edition*. CRC Press, 2001.

[2] L. L. van Dierendonck, J. Zahradnik, and V. Linek, "Loop Venturi Reactor A Feasible Alternative to Stirred Tank Reactors?," *Ind. Eng. Chem. Res.*, vol. 37, no. 3, pp. 734–738, 1998.

[3] M. Di Serio, R. Tesser, and E. Santacesaria, "Comparison of Different Reactor Types Used in the Manufacture of Ethoxylated, Propoxylated Products," *Ind. Eng. Chem. Res.*, vol. 44, no. 25, pp. 9482–9489, Dezember 2005.

568871 Process Intensification of Propane Dehydrogenation Using Microporous Silica Membranes

Propylene is an important intermediate chemical for production of petrochemicals such as polypropylene and acrylonitrile. Propylene and other light olefins are mainly produced by light hydrocarbon fractions such as steam cracking and fluid catalytic cracking. The equilibrium limit exists in traditional packed bed reactors (PBR), thus packed bed membrane reactors (PBMR) became an important subject to explore because they can combine chemical reaction and separation in one step. Due to selective removal of one of the products (H_2), PBMR helps in shifting the equilibrium towards the forward reaction, and can overcome the equilibrium limitations in PBR. In this work, we developed microporous silica membranes modified with atomic layer deposition (ALD) for effective H_2 separation from hydrocarbons. The ALD-modified silica membrane was used in PBMR for PDH reaction in combination with Pt/Al_2O_3 catalyst. Moreover, one-dimensional plug-flow reactor (PFR) model was established to investigate the effect of membrane properties and operating conditions on PDH reaction systems.

570440 Scalable Thin-Layer Membrane Reactor for Heterogeneous and Homogeneous Catalytic Gas–Liquid Reactions

Catalytic gas-liquid reactions have potential as environmentally benign methods for organic synthesis, particularly hydrogenation and oxidation reactions. However, safety and scalability are concerns in the application of gas-liquid reactions. In this work, we develop and demonstrate a scalable, sustainable, and safe thin-layer membrane reactor for heterogeneous Pd-catalyzed hydrogenations and homogenous Cu(I)/TEMPO alcohol oxidations. The implementation of a Teflon amorphous fluoroplastic (AF) membrane and porous carbon cloth in the membrane reactor provides sufficient gas-liquid mass transfer to afford superior performance compared to conventional packed-bed or trickle-bed reactors. The membrane separates the gas from the liquid, which avoids the formation of explosive mixtures for oxygenation reactions and simplifies the two-phase hydrodynamics to facilitate scale-up by stacking modules, while significantly reducing gas consumption. In addition, 3-dimensional simulations deliver insights into the mass transfer and hydrodynamic behavior to inform optimal membrane reactor design and operation.

577756 Plug Flow Membrane Reactor for Esterification of Acetic Acid with Mor-Type Zeolite Membrane

While esterification is important for chemical production, the attainable level of conversion is always thermodynamically limited. Thus, the product of esterification is a mixture of organic acid, alcohol, ester, and water. Dehydration membrane reactor has been drawn attention as a novel energy-saving process for esterification for recent decades. When a dehydration membrane is adopted to remove water from the reaction system, the attainable level of conversion will be improved beyond that limited by thermodynamics in conventional reactor. Consequently, one can expect that energy consumption for purification process after reaction is reduced as well as the improvement of one-path yield of products.

We previously reported MOR-type zeolite membrane for dehydration and pointed out that our MOR membrane had high chemical resistance for organic acid.[1] In this study, we prepared applied MOR membrane to a plug flow reactor for the esterification of acetic acid.

MOR membranes were synthesized by a seed-assisted crystallization method on the outer surface of porous tubular a-alumina support (Noritake, i.d. = 7 mm, o.d. = 10 mm) of which the average pore size of the top layer was ca. 150 nm. The outer surface of support was seeded by a dip-coating method using a colloidal suspension of MOR seed crystals.

The seeded support was placed vertically in a Teflon-lined stainless steel autoclave and filled with a reaction mixture. The reaction mixture had a molar composition of $10Na_2O:0.15Al_2O_3:36SiO_2:960H_2O$ and was prepared by mixing colloidal silica (ST-S, Nissan Chemical Ind. Ltd.), sodium aluminate (Kanto Chemical Co. Inc.), sodium hydroxide (>97 wt%, Kanto Chemical Co. Inc.), and deionized water. The reaction mixture was aged at 323 K for 4 h prior to use. Hydrothermal crystallization was carried out at 453 K for 7 h in an autoclave. After the crystallization, the autoclave was quenched with water. The synthesized membrane was washed in boiling water and dried over night at 383 K.

Dehydration performance of MOR membrane prepared was evaluated by the separation test for an equimolar mixture of water and acetic acid in a pervaporation mode at 343 K. The permeation side, the inner side of tubular membrane, was evacuated by using a vacuum pump. The permeate was condensed by a cold trap and was analysed for its composition by using GC-TCD.

The esterification of acetic acid was carried out in a plug flow membrane reactor as follows. The equimolar mixture of acetic acid and ethanol was fed to the outer surface of tubular membrane. Amberlyst-15 was packed on the outer side of membrane as catalyst. Temperature and pressure were adjusted at 403 K and 1.0 MPa, respectively. The permeate was analyzed for its composition. The yields of ethyl acetate produced in the plug flow reactor with and without MOR membrane were compared.

As a result of separation test for acetic acid and water, MOR membrane showed great dehydration performance. The separation factor for water/acetic acid was above 139 with its permeance of 1.65×10^{-7} mol m^{-2} s^{-1} Pa^{-1}. Separation and permeation performance were stable at least up to 8 h, suggesting that MOR zeolite is one of the feasible membrane materials for dehydration under acidic conditions.

Whereas the yield of ethyl acetate in the plug flow membrane reactor without MOR membrane was thermodynamically limited at 66 %, its yield was improved up to 72 % with the membrane in the reactor tested. Selective removal of water by MOR membrane contributed to the enhancement of ethyl acetate yield.

(1) G. Li, E. Kikuchi, M. Matsukata, The control of phase and orientation in zeolite membranes by the secondary growth method, *Microporous Mesoporous Mater.* 62 (2003) 211-220

Acknowledgment

This work was partially supported by NEDO (New Energy and Industrial Technology Development Organization), Japan.

563345 Continuous-Flow Synthesis of the Azo Pigment Yellow 14 Using a Three-Stream Feeding Micromixer

Abstract:

Organic coloured pigment is one of the most important chemical fines. The demand for organic coloured pigments continuous to grow with the growing use in the field of colour printing、cosmetics、coatings、plastics and so on. Among the organic coloured pigments, the most well-known class is the azo pigment. One example is azo pigment yellow 14, which is a typical bisazo pigment. The yellow pigment is usually the last component of the printing process[1] and therefore, they often need to have some special properties, such as a high transparency、a high strength of colour and a high tinctorial power[1]. The colour properties of azo pigments are connected with the particle size and the particle size distribution. They are affected by the production process [1].

In industry, the azo pigments are often synthesized by batch process using stirred tanks with volumes varied from 20 m^3 to 80 m^3[1]. However, batch processes are possible to lead to a relatively board size and a low conversion of the coupling component, which seriously affects the quality of the azo pigments. For the synthesis process of azo pigment yellow 14, it includes double diazotization and double coupling reaction, which need an accurate regulation of the pH during reaction process. Therefore, a better mixing effect and a more precise adjustment of synthesis process are needed. In last 20 years, the microreactor technology has been widely used in both industry and academia[2]. Compared with traditional batch vessels, the microreactor technology has many advantages, such as better mass and thermal transfer, more accurate control parameters[3].

The first successful synthesis of azo pigments in microfluidic devices was reported by Wille et al.[4] The azo pigments showed improvements in color properties compared with batch processes. Then, the synthesis of Pigment Yellow 12 in a micromixer apparatus was developed by Pennemann et al.[1] Using a Micromixer-based process, they improved the properties of the azo pigment yellow 12. However, the azo pigments synthesized above were not sensitive to the changing of the pH during reaction process compared with the azo pigments yellow 14.

In this work, a three-stream feeding process for the synthesis of azo pigment was considered to control the pH of the coupling reaction process accurately. The acetic acid/sodium acetate buffer solution was pumped into the micromixer as a separate stream which was different from the two-stream feeding process and batch process. The pH of the reaction process was not necessary to be monitored at any time if the pH of the buffer solution was controlled. Furthermore, a three inlets microreactor for easy scale-up study was designed to carry out the coupling reaction. Then, the effects of pH of the buffer solution、reactants flow rates and reaction temperature on the purity and the particle size distribution of the azo pigments were discussed. Finally, the hue and transparency of the products compared with pigments standard (commercial products, SHENYANG RESEARCH INSTITUTE OF CHEMICAL INOUSTRY) were also studied. The results show that under optimal conditions, a higher purity (94.54%) of the pigment can be obtained using three-stream feeding process than that using two-stream feeding process (83.34%) and a monomodal particle distribution can be obtained compared the batch process getting a bimodal distribution. According to the coordinates of CIELab colour space (L*, a*, b*) evaluated for azo pigments, it illustrated that the red hue and the blue hue of the azo pigment products were larger compared with pigment yellow 14 standard. The lightness L* of the pigments synthesized by three-stream feeding process was larger than that by two-stream feeding process. The transparency was increased by 98% compared to a commercially available yellow 14. Overall, the micromixer-based three-stream feeding process led to a significant improvement of the azo pigment yellow 14 process, which had a good industrial application value.

References:

[1]Pennemann, S. Forster, Kinkel, Improvement of Dye Properties of theAzo Pigment Yellow 12 Using a Micromixer-Based Process. Org. Process Res. Dev. 9 (2005) 188-192.

[2]J.S. Zhang, K. Wang, Y. C. Lu, G. S. Luo, Characterization and modeling of micromixing performance in micropore dispersion reactors. Chem. Eng. Prog. 49 (2010) 740-747.

[3]Z. Q. Yu, X. Ye, Q. L. Xu, X. X. Xie, H. Dong, W. Su, A fully continuous-flow process for the synthesis of p-Cresol:Impurity Analysis and process optimization. Org. Process Res. Dev. 21(2017) 1644-1652.

[4]C. Wille, V.Autze, H. Kim, U. Nickel, S. Oberbeck, T. Schwalbe, L. Unverdorben, In Proceedings of the 6th International Conference on Microreaction Technology, IMRET 6, March 11-14,2002, New Orleans, LA, U.S.A.; pp 7-17.

AUTHOR INDEX

AUTHOR INDEX

AUTHOR INDEX

AUTHOR INDEX

AUTHOR INDEX

AUTHOR INDEX

AIChE
120 Wall Street, FL 23
New York, NY 10005-4020

ISBN 978-1-7138-0535-9

9 781713 805359

CURRAN ASSOCIATES INC.
proceedings.com